AF540001

DATA HANDLING IN ANALYTICAL CHEMISTRY

DATA HANDLING IN ANALYTICAL CHEMISTRY

By

Radhika Garg

2016

SBS Publishers & Distributors Pvt. Ltd.
New Delhi

ISBN 13 : 9789380090955

First Published in 2016

Published by:

SBS PUBLISHERS & DISTRIBUTORS PVT. LTD.

2/9, Ground Floor, Ansari Road, Darya Ganj,

New Delhi - 110002,

INDIA

Tel: 0091.11.23289119 / 41563911

Email: mail@sbspublishers.com

www.sbspublishers.com

Preface

Analytical chemistry is a scientific discipline that develops and applies methods, instruments, and strategies to obtain information on the composition and nature of matter in space and time, as well as on the value of these measurements, i.e., their uncertainty, validation, and/or traceability to fundamental standards. Analytical chemistry is a quantitative science. Whether determining the concentration of a species, evaluating an equilibrium constant, measuring a reaction rate, or drawing a correlation between a compound's structure and its reactivity, analytical chemists engage in measuring important chemical things. Data handling includes the collection, presentation, interpretation and evaluation of data from both primary and secondary sources. The quality of the data is extremely important when assessing the strengths and limitations of models and theories. Data must be accurate, reliable and repeatable if it is going to be used in this way. Scientists often present their findings through various kinds of visual representation. Scientists need a standard format to communicate significance of experimental numerical data.

Editor

Contents

Chapter 1

REMOTE REAL-TIME MONITORING OF SUBSURFACE LANDFILL GAS MIGRATION

Cormac Fay, Aiden R. Doherty, Stephen Beirne, Fiachra Collins, Colum Foley, John Healy, Breda M. Kiernan Hyowon Lee, Damien Maher, Dylan Orpen, Thomas Phelan, Zhengwei Qiu, Kirk Zhang, Cathal Gurrin, Brian Corcoran, Noel E. O'Connor, Alan F. Smeaton * and Dermot Diamond *

CLARITY, Centre for Sensor Web Technologies, Dublin City University, Glasnevin, Dublin 9, Ireland;

E-Mails: cormac.fay@dcu.ie (C.F.); aiden.doherty@dcu.ie (A.R.D.); stephen.beirne@dcu.ie (S.B.); fiachra.collins@dcu.ie (F.C.); Colum.Foley@computing.dcu.ie (C.F.); john.healy2@mail.dcu.ie (J.H.); Breda.Kiernan@dcu.ie (B.M.K.); hlee@computing.dcu.ie (H.L.); damien.maher@dcu.ie (D.M.); dylan.orpen@dcu.ie (D.O.); thomas.phelan4@mail.dcu.ie (T.P.); zhengwei.qui@computing.dcu.ie (Z.Q.); ke.zhang@ucd.ie (K.Z.); cgurrin@computing.dcu.ie (C.G.); Brian.Corcoran@dcu.ie (B.C.); Noel.OConnor@dcu.ie (N.E.O.)

⋆ Author to whom correspondence should be addressed; E-Mails: alan.smeaton@dcu.ie (A.F.S.); dermot.diamond@dcu.ie (D.D.).

ABSTRACT

The cost of monitoring greenhouse gas emissions from landfill sites is of major concern for regulatory authorities. The current monitoring procedure is recognised as labour intensive, requiring agency inspectors to physically travel to perimeter borehole wells in rough terrain and manually measure gas concentration levels with expensive hand-held instrumentation. In this article we present a cost-effective and efficient system for remotely monitoring landfill subsurface migration of methane and carbon dioxide concentration levels. Based purely on an autonomous sensing architecture, the proposed sensing platform was capable of performing complex analytical measurements *in situ* and successfully communicating the data remotely to a cloud database. A web tool was developed to present the sensed data to relevant stakeholders. We report our experiences in deploying such an approach in the field over a period of approximately 16 months.

Keywords: environmental monitoring; greenhouse gases; chemistry; landfill; carbon dioxide; methane; sensor networks; sensor data management

INTRODUCTION

Global Environment

Global warming is recognised as a serious worldwide challenge. The Intergovernmental Panel on Climate Change (IPCC) fourth report states that the warming of our climate is evident and that human activities are very likely the cause through the emission of substantial amounts of greenhouse gases into the atmosphere [1]. In this article we focus on greenhouse gases (methane and carbon dioxide) emitted from the decomposition of biodegradable waste at landfill sites.

Chemical Sensing and Information Retrieval from the Environment

It is well documented that the quality of our environment is determined by its chemistry, and imbalances in a wide variety of parameters can have a drastic effect on air and water quality, leading to increases in the incidence of respiratory diseases and cancers, amongst others [2–5]. Our basic sensing capabilities are limited and only allow us to observe the aftermath of the effects that these substances ultimately cause. The ability to harvest chemical information from our environment can provide the means to enforce preventative measures and/or to provide early warning. Hence, chemical sensing is on the increase and is very much encouraged by local and global governmental legislation, such as the Water Framework Directive [6], Kyoto protocol [7], Climate Change Act [8], GlobalWarming Solutions Act [9].

However, target-specific environmental chemical sensing is not easy to achieve due to the many problems of integrating chemical sensors into practical, long term sensing platforms [10]. For instance, to realise a fully functional chemical sensing system, the designers must face a multitude of multidisciplinary issues such as: reduction of the sensing surface, drift, cross sensitivity, bio-fouling, chemo-electronic transducer, power consumption, elemental robustness, security, vandalism/damage, autonomous control, successful and secure delivery of data to stakeholders/authorities.

The concept of Internet Scale Sensing (ISS) for chemical sensing is well known but it should be noted that although new and emerging techniques from the digital and chemical worlds continue to progress, a critical missing element is "the gateway" linking these two realms [11]. A key goal for autonomous environmental monitoring is in providing the ability to easily access that data by relevant stakeholders such as environmental enforcement agencies. It is believed that the use of the *"cloud computing"* concept, where data is stored on a server and is always available via the Internet, will be the means to access harvested environmental data. This data can be viewed across a wide variety of computing/software platforms, e.g., internet browsers, iPhones, iPads, smart phones, *etc.* The final vital ingredient behind the Internet Scale Sensing vision is in turning the massive volume of raw sensor data into meaningful information. Some have suggested that this process will be achieved via signal processing with event detection models [12]. After those events have been identified, software outlier detection algorithms can then be employed to identify the events of most interest and subsequently alert the relevant stakeholder(s).

Landfill Emissions and the Current Monitoring Standard

Waste activities account for approximately 5% of the global greenhouse gas (GHG) emissions and of all the waste management methods in use, land-filling is by far the most common [13,14]. The major components of landfill gas (LFG) are typically carbon dioxide (CO_2: 40%–60%) and methane (CH_4: 40%–60%) from the decomposition of biodegradable waste [15,16]. During landfill cell development, and before introduction of the waste body, the surrounding soil is covered with a mineral layer in conjunction with a geosynthetic liner. Further to this, a network of perforated pipes is implemented in order to fully extract the landfill gas being produced. After capping, the gas is subsequently treated and/or disposed of in a safe manner through flaring or venting [17]. Following decomposition of the waste, landfill gas generation can begin as early as 6–12 months after capping and typically continue for a further 20–50 years [18] and even as long as several hundred years in some cases [19]. In addition, serious health issues have been linked to landfill site proximity [20–23]. It is clear that limits must be put in place on the gas concentrations emissions. As a result, the waste license for landfill sites, from the Irish Environmental Protection Agency, states that concentration levels (measured from perimeter borehole wells) must not exceed 1% for methane and 1.5% for carbon dioxide [24] (it is noteworthy that the lower explosive limit for methane is 5% v/v in air). To comply with these levels, the site's flare is designed to dispose of harmful gases (e.g., methane) through the gas extraction network. If a problem is identified (through manual monthly monitoring) such as an increasing gas concentrations, the flare is then set to increase its production.

Arising from legislative enforcement, emission levels from landfill sites require continuous monitoring including sub surface migration of gases [25]. Monitoring is required to take place at an average frequency of once per month—and can even be as infrequent as 4 times per year in some cases—by using a hand-held gas analyser to monitor samples extracted from the top of perimeter borehole wells for a duration of approximately one minute. Subsequently, these levels are manually reported to national environmental protection agencies by the license holders via email/phone. If in breach of permitted gas concentration limits, fines are subsequently levied. However elements of human error and variability across operators can provide problems. Also, in extreme situations, this process is vulnerable to manipulation because inspectors must announce their visits beforehand, allowing for the possibility of high concentration gases to be vented to the atmosphere prior to inspection.

Most of the available gas analysers suitable for landfill gas monitoring are handheld based. At present, a majority of the borehole wells in Ireland are monitored using the GA2000 (Geotech) as standard.

However other available products offer similar functionality including GA90/GEM500/GEM2000 (Landtec), G3 LMSxi (Ashtead) and Gas-Tec (AFC). In the same manner, specialised instruments exist that offer a continuous monitoring feature suitable for landfill site monitoring including IR600 (Hitech Instruments), LFG2003 (Liberty Engineering), Ultramate 23 (Siemens), AEMS (Geotech) *etc.* However, many of these systems are heavy and relatively expensive, require specialised personnel to maintain and install, demand high power (most requiring mains power) and are aimed at monitoring at the flare only *i.e.,* giving a global reading for the entire extraction system network. Hence, there is a need for portable, low power, low cost instrumentation to continually monitor localised areas and ultimately replace the manual monitoring tasks.

Chemical Sensing of Greenhouse Gases from Landfill Sites

Given this background we feel that an automated monitoring technique is essential to reduce the cost of manual sampling, improve the reliability of the gas concentration levels reported to governing authorities and ultimately provide a means to better identify elevated GHG emissions, which may lead to additional steps to reduce those emissions. In this paper we apply the template of internet scale sensing and detail a prototype end-to-end sensing model for landfill emission monitoring, which has been deployed in field studies over a time period of 16 months across three different locations. We have collected 432,540 sensor readings with 2,403 remote communications back to our web database. Specifically, we describe the gateway platform (Section 2), show approximately one year of harvested data from an extended *in-situ* trial on an active landfill site (Section 4.3) and finally, we describe potential improvements which can be implemented from this development phase to further reduce costs (Section 4.6).

INTERNET SCALE SENSING FOR LANDFILL EMISSION MONITORING

Sensing Model

To achieve our goals, we applied the concept of internet scale sensing (ISS) and expanded it to suit a realised sensing model for real-time landfill migration monitoring via a live web-site. Figure 1 provides a visual representation of our end-to-end sensing model. Firstly, the physical/chemical gas sensors are exposed to the sample landfill gas and generate an electrical signasl that are proportional to the concentration of the constituents within the target gas sample. These signal lines are conditioned by the gateway platform and digitised by the system's micro-controller. At this point, the sensor data are stored on-board and transmitted via the GSM network to a base station where the data are then forwarded to a web-enabled sensor network server. After storage on a relational database, the data are visually presented to web users and governing authorities by means of an easy to interpret web page. The following sub-sections gives a more detailed explanation of these processes.

Gas Sensors

The system was equipped with a humidity sensor (Honeywell HIH-4000-001), linear range 0 to 100% RH, and a temperature sensor (a 10 k thermistor—Thermometrics DKF103N5) with an operational response range of −40⁰C to +250⁰C. Both temperature and humidity measurements are critical for providing background information when analysing gas samples, especially during developmental stages and for correlation with environmental artifacts that may give rise to erroneous signals from the chemical sensors. More importantly, the system was also equipped with two chemical gas sensors; for CH_4 and CO_2 detection. As discussed earlier (Section 1.3), landfill gas is composed mainly (greater than 95% [16]) of CH4 and CO2, thus these were the principle sensing targets. Although the permitted concentration limits of both gases are low (see Section 1.3), it was found through previous developmental phases and field validation trials that measured readings were up to 10 times that of the allowed regulated range [26].

As a result, both sensors (NDIR based) were sourced from Dynament Ltd. (www.dynament.com) with a custom range of 0%–20% to suit problematic landfill sites. These sensors were designed to be self compensating to both temperature and humidity in the ranges of −20 ◦C to +50 ◦C and 0% to 95% RH respectively [27,28].

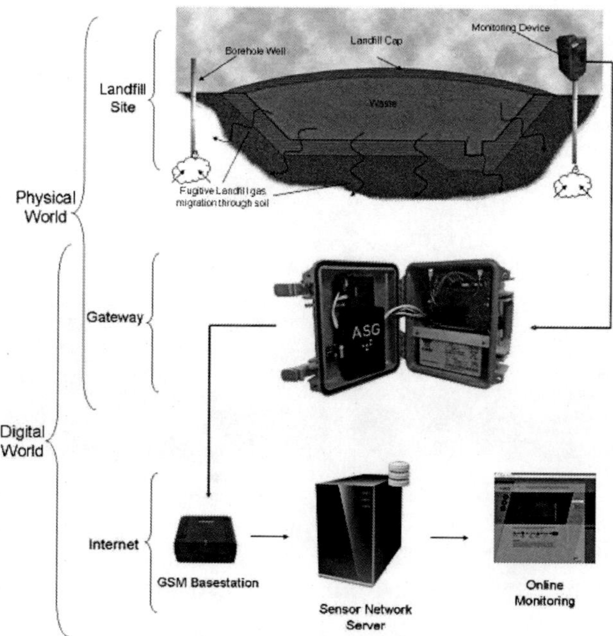

Figure 1. Visual representation of the landfill gas sensing model from the de-

vice placed in the field to a web-based visualisation user interface. The model shows the progression of chemical sensed data from the physical world to the digital world by means of a gateway platform.

Gateway Platform

System Components

All system components were housed within a robust case (Peli Case 1300). An internal volume of 25.1 (L) × 17.8 (W) × 15.5 (H) cm allowed sufficient space for all system elements to be securely packaged and for a systematic layout of components. The case's lid held the most accessible and frequently used system elements employed by users during system development and deployment, *i.e.*, control and communications. Conversely, the sensors and actuators were positioned within the case's base for secure and fixed positioning while allowing sufficient room for gas tubing and electrical connections. Figure 2 shows the layout views of the system's components for control/communication and sensors/actuators respectively.

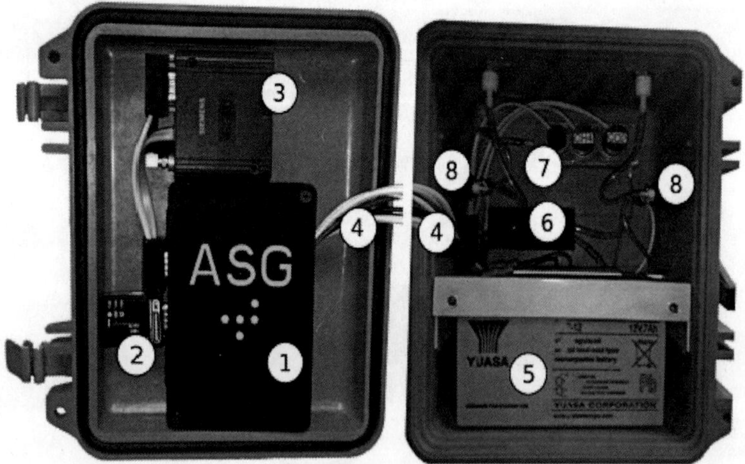

Figure 2. Component layout of the gateway platform. (1) control system, (2) bluetooth module, (3) GSM module, (4) signal and actuation control lines, (5) power source, (6) extraction air pump, (7) gas chamber, (8) flow selection valves.

Micro-controller Board

A custom, in-house designed and assembled PCB (Printed Circuit Board—manufactured by Beta LAYOUT Ltd.) was developed to suit the requirements of this project. At its core lay an MSP430F449 (Texas Instruments) which controlled component power switching, timing, signal conversion, data handling, storage, communications *etc. i.e.*, all processes necessary to achieve full autonomy of the system.

The board was equipped with ten switch-able power ports (8 × PFETs and 2 × NFETs). Each port's power supply was capable of being manually selected (via on board jumper pins) from 3 separate voltage supplies *i.e.*, 3V3 regulator (Texas Instruments, LP2985A), 5 V regulator (National Semiconductor, LP2992IM5) and 12 V (main battery).

Short Range Communications

A miniature bluetooth module (LM Technologies LM048) was included for local, short range communications. This feature allowed users to interface with the system without exposing the electrical components/connections such as the microcontroller to external conditions, such as rain. The module communicated to the microcontroller through a DE9 connection, via an RS232 transceiver (Maxim, MAX3232CSE) and finally to one of its UART channels. Power to the bluetooth module was applied by means of an external, weatherproof (IP67 rated) switch and using the 5 V voltage regulator.

Long Range Communications

Remote reporting of landfill gas concentrations to a local base station was achieved by means of a GSMmodule (SiemensMC35iT). This allowed stand-alone developmental systems to be deployed in very remote locations while still being able to communicate data back to stakeholders. Power was supplied via three PFET power switches. Communications were achieved in a similar fashion as with the bluetooth module with the exception of using the microcontroller's secondary UART channel.

Signal and Actuation Control Lines

Wiring for switching power to components was of single core and insulated (RS 140420). Sensor signal lines were connected using shielded wiring (RS 1643740).

Power Source

A low cost, rechargeable and high capacity 12 V battery supplied power to the entire system (YUASA-NP712). Note: a 2× AAA battery pack supplied power to the microcontroller (behind a diode) for an uninterrupted power supply to the microcontroller which allowed the main battery to be changed with no loss of RAM e.g., the real time clock.

Extraction Pump

An air pump (SKC Grab Air 222-2301) with closed circuit input/output ports allowed sample gas to flow through the sample chamber at a rate of 0.6 L/min. This was actuated by a single microcontroller I/O port and one NFET with power from a 5 V regulator.

Gas Chamber

A custom designed gas chamber accommodated all sensors and two port connections. It was fabricated using a Dimension SST 768 rapid prototyper and uniformly sealed using a combination of MEK and silicon coupled with O-rings for sensor access.

Flow Selection Valves

In order to select the supply source and target exhaust flows, we equipped the system with two 3/2 way latching pneumatic control valves (Lee Products Ltd. LHLA0531211H).

SAMPLING PROCEDURE

As discussed earlier, the existing legislative monitoring procedure (for determining landfill gas emissions from perimeter borehole wells) calls for a sampling frequency of once per month, extracting gas for circa 1 minute using a hand-held instrument and venting to atmosphere (Section 1.3). In this study, we increased this sampling frequency 60 fold. However, this immediately meant that much more greenhouse gases would be emitted to atmosphere when following the existing sampling procedure; a contamination that this device was ultimately expected to reduce. To address this issue, a previous study was conducted to investigate a recycling technique whereby the sampled gas was exhausted back into the borehole well (at a different depth) instead of venting to atmosphere [26]. The findings of the study showed that this recycling technique (of landfill gas recycling back to the borehole well) did not affect the gas composition measurements when compared with venting to atmosphere. Moreover, the same study discovered that when multiple borehole wells of various head space depths were sampled, the longest time to achieve a steady state measurement was circa 2 minutes. To allow for an appropriate settling time, a 3 minutes monitoring duration was chosen. As a result, the automated monitoring time was divided into 3 separate procedures (*baseline, sample* and *purge*) where each was monitored for a duration of 3 minutes and sampled at 3 seconds intervals. The device's air flow control system is illustrated in Figure 3.

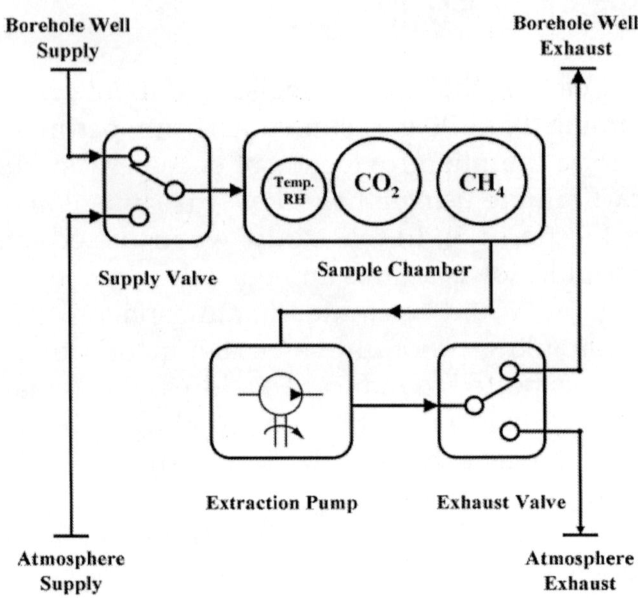

Figure 3. Schematic flow diagram illustrating the gas flow control system. The flow control valves allow the system to be switched between sampling mode (from 'BoreholeWell Supply Supply' to the 'Borehole Well Exhaust') and baseline and purge modes (from 'Atmosphere Supply' to the 'Atmosphere Exhaust').

Firstly, for the *baseline* procedure, the "Supply Valve" was switched to the "Atmosphere Supply", and the "Exhaust Valve" to the 'Atmosphere Exhaust' settings. The motivation for this step was threefold. The primary reason was to check that the sensors were in fact powered up, *i.e.*, a valid measurement gave a value of no less than the sensors standard offset of 0.4 V at 0% v/v (if unpowered the sensors report an electrical potential of 0 V). Coupled with this, it allowed sufficient time for the IR sensors to warm up and stabilize; a necessary step for accurate measurements. Furthermore, it was also ensured that no residual landfill gas was present in the chamber from previous measurement cycles. Subsequently, the *sample* procedure took place whereby both valves were toggled from the *baseline* step so that landfill gas was extracted from the "Borehole Well Supply" and exhausted to the "Borehole Well Exhaust". Finally, the valves were again toggled to the state used for the *baseline* procedure. The following *purge* procedure was used to remove the landfill gas from the gas chamber.

SIGNAL AND DATA FLOW

A total of five signal lines were connected to the microcontroller's analogue to digital converter (ADC) channels. As a rule, the ADC channels may only digitise voltage levels between 0 V and 3.3 V. The dynamic range of each analogue signal line was first conditioned to lie within the ADC measurement range by means of appropriate signal conditioning circuitry. Once conditioned, the sensor outputs were digitised, via the ADC channels, at 3 seconds intervals over the full 9 minutes sampling procedure. Along with the sensor data, a single battery reading was saved as part of the data set, along with a timestamp according to the on board real time clock (set to GMT). The full data set was retained in RAM before storage or reporting back to the stakeholder. Following this, the data was saved as raw ADC values on two separate 2 Mbit flash chips (Numonyx M25P20-VMN6P) allowing a total of 131.072 Kbytes to be stored over 8 separate sectors. The data were arranged in a structured format to utilise all data bits in each byte. The first, second and third sectors on each chip were assigned to store the data harvested during the baseline, sample and purge procedures respectively. By the same token, the fourth sectors were kept in reserve for additional sensor data overflow or system specific settings. As a result, the system was able to save up to two years of separate trial data assuming a sampling frequency of twice per day.

COMMUNICATIONS

As environmental devices are often placed in remote locations, one of the only readily available methods of wirelessly transmitting sensed data to stakeholders is to take advantage of the GSMnetwork's national coverage. The system's GSM module was powered via three of the eight positive switching power channels for mains power, wake up and shut down triggering. After harvesting and storing landfill gas data from a monitoring cycle, a statistical representation was compiled (average, max, min) and sent to the Sensor Network Server (SNS). SMS messaging is used at present with optional use of the module's GPRS feature; however it was found to be more effective to use SMS for maintenance and continuous power costs. For complete dataset retrieval, one must interface with the device and download stored data from the on-board flash memory chips. Initially, this was achieved in the same manner as in a laboratory setting, *i.e.*, a direct wired connection to a laptop PC. However, this method is highly undesirable (*i.e.*, to open the system *in situ*) as it left the system vulnerable to such conditions as rain, wind and biofouling, potentially causing damage such as electrical shorting. As a result, the system was equipped with an external power control switch (IP67 rated) for the Bluetooth module. Thus, full system control was accessible for diagnosis of individual components, setting the system clock and retrieval of data without breaching the system's environmental seal.

Sensor Network Server

Base Station and Database Interface

The GSM base station (Siemens MC35iT) was connected to a server PC via a RS232 interface (Rotronic UC232A) and powered by a standard 12V power adapter (MasterplugMVA1200-MP).Written in the Java programming language and using the Javax.comm package, a custom written application, denoted here as the GSM Database Interface (GDI), provided the functionality to progress data from new Landfill SMS texts to a secure MySQL database. Figure 4 depicts the overall process of advancing the data from the GSM base station to the primary environmental database.

At first, the base station was placed into "SMS Alert Mode" where, upon receipt of a new SMS, a string of characters was sent to the server indicating which slot the new text was stored within the GSM's memory. Once the GDI was alerted to a new packet, it automatically parsed the alert string and extracted the SMS memory storage address on the base station. Next, the GDI requested and received the full SMS text located at the appropriate memory address. After parsing the new text header, the sender's phone number was used as the unit identification factor. At this point the GDI connected to a "registry database" where it received all relevant information related to the remote device such as type, unit number and calibration settings based on the sender's phone number. Next, the text body was parsed. All reported measurements were extracted and automatically placed on the database using standard SQL statements. Also, inline conversions of ADC to CO_2, CH_4 and battery levels occurred at this point using previously acquired calibration data (Section 3.1) in the lab. Finally, the text was deleted from the base station as the storage capacity of the SIM card was limited.

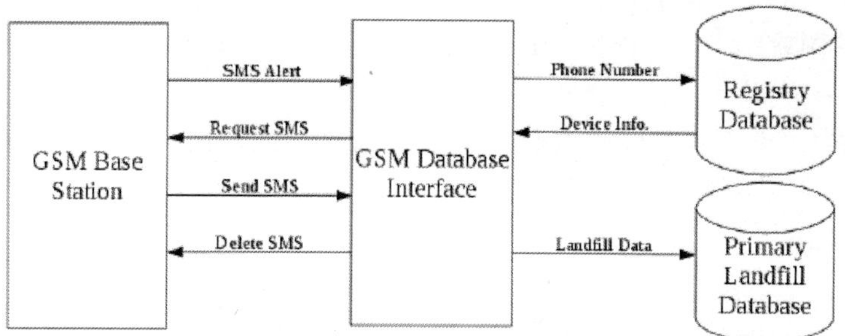

Figure 4. Block diagram showing the interactions between the GSM base station, the GSM database interface and the relational databases. The remotely reported data is received by the GSM base station where, through a number of programming stages, the data is stored on the primary landfill database.

Data Organization

Once data was transmitted via the GSM network to a central base-station, the data was stored in a relational database which can manage data from multiple landfill locations. This relational database was constantly available on the Internet which thus fulfils our mission of data accessibility, and also it was stored in a replicated database setup; meaning that there was multiple copies of the data, thus ensuring data redundancy. Figure 5 provides an overview of our end-to-end system, from sensing our environment through to informing users of the relevant levels of CO2 and CH4 in their local landfill site.

Data Presentation

The data were accessible by relevant stakeholders via an intuitive web interface in which the data were displayed graphically. Figure 6 shows the designed Silverlight web interface. Initially the user is presented with a list of trials (both live and historical) via a combo box on the right hand side of the screen. After selection of a relevant trial the user can click on the relevant radio button on the right hand side to view either the CO2 or CH4 values, which are displayed as a bar chart, with yellow-green bars indicating values within recommended limits and orange bars indicating that the limits have been exceeded. The user can change the visualisation unit to daily, monthly, and yearly by clicking on the option buttons, and arrow buttons are provided to slide the time period to the previous or next time span.

Finally, below the main display area a brief description of how the information is collected is

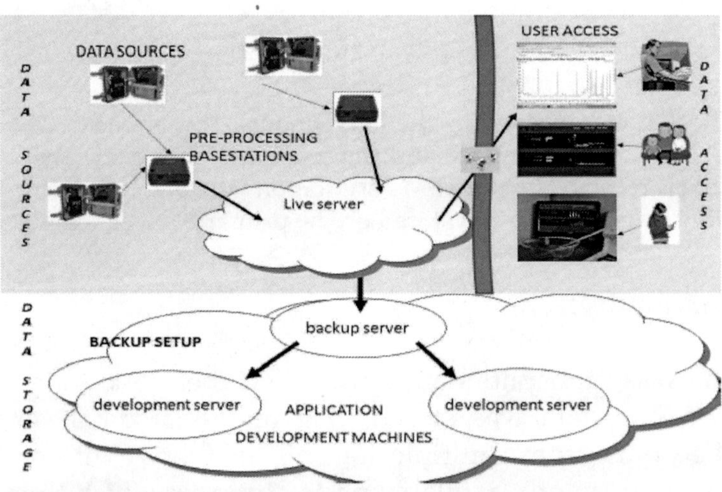

Figure 5. Overview of Data Storage, Backup and Presentation. Multiple landfill sensors can upload data to a single cell base-station. Thereafter these base-stations upload data to a central server, which is also backed up. Finally this data is available via the Internet for end users to access.

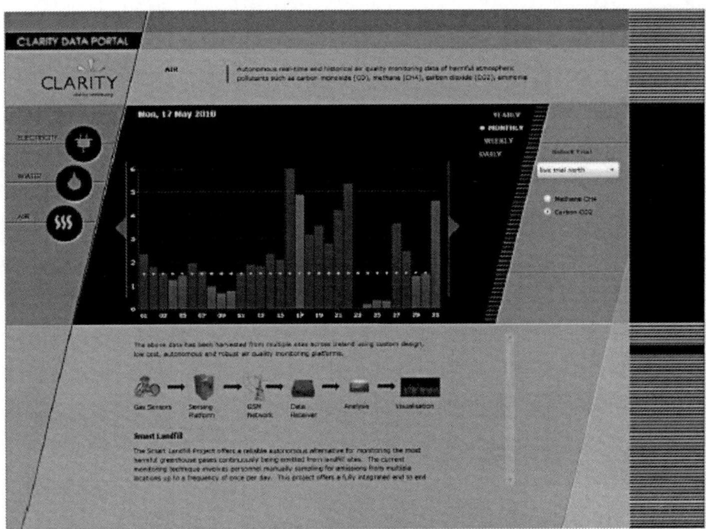

Figure 6. Sensor Data Portal: Our web application which allows relevant stakeholders to easily view in real-time the air quality (CO_2 & CH_4) data from landfill sites. The website can be viewed at http://clarityapp.ucd.ie/~sensorportal/.

METHODOLOGY

In this section we now describe the various procedures and experimental setups that were utilised to evaluate our system. Firstly, it was necessary to thoroughly calibrate the relevant sensors in a laboratory environment. Most importantly, we then describe our protocol for field deployment for a period of over one year. Finally, we review the protocol to carry out post-analysis software optimisation simulations on the array of data gathered in the field in the previous year.

Calibration of Chemical Sensors

Before any deployment of the system could take place it was first essential to calibrate the chemical sensors. At first, the system was setup in the same configuration for field trials *i.e.*, when sampling and exhausting to/from a borehole well except the supply was connected to source gases and the exhaust to a fume hood extraction system. Next, the microcontroller executed a pre-programmed calibration routine where all the sensors were powered on and the digitised ADC values were continually output (in an endless loop at a frequency of 0.33 Hz) to a laptop computer via a serial port and captured using Microsoft Hyperterminal (a command line interface communications utility). Each sensor was calibrated against source gases (CO2/CH4 supplied by Scott Specialty Gases) at concentrations of 0% to 50% with a nitrogen balance. Coupled with a dilution of ambient air, sourced using an air compressor (Werther International 42040 100/50), and managed with mass flow controllers (Cole Parmer YO-32708-26), various gas concentrations were achieved for calibration of the CO_2 and CH_4 sensors. A flow rate of 0.6 L/min (matching the flow rate of the system's air pump) was ensured using a standard flow meter. Furthermore, a GA2000 Plus device (the current landfill gas monitoring standard hand-held gas analyser) was used as reference and verification of both gas concentrations and flow rate. A ten point calibration plot for each IR sensor (CO_2 and CH_4), was achieved (Section 4.1).

Power Usage

The system was programmed to autonomously wake up from a low power mode every 12 hours (changed to every 6 hours subsequently), perform a monitoring (analyse gas composition and report) cycle and subsequently return to its low power state. As this system was placed at a remote location, it was desirable to establish how long the system would remain operational using the existing power source (7 Ah lead acid battery). Consequently, current consumption analysis was performed using a high end multi meter (Keithly 2700) capable of sampling at a frequency of 60 Hz with a resolution of 9 decimal places. The landfill system's power source was connected in series with the multi meter and configured for a typical monitoring cycle – see Section 4.2 for analysis.

System Deployments

While the overall sensing model was in the final stages of development, including the implementation of 3 deployable platforms, a parallel effort was undertaken to find a suitable location for our first deployment. Ideally we wanted a site that would allow us access, be safe from possible vandalism and have an eventful borehole well to monitor. As many landfill sites in Ireland are privately owned, our first criteria was not easily met. After much evaluation, we located a site (closed to public access) where the personnel were very accommodating and enthusiastic, as they had a problematic well that needed continuous monitoring (*location A*). After the site personnel experienced the remote sampling advantages of the system, we were asked to monitor a second well on the same site (*location B*). Some time afterwards, our national environmental enforcement agency requested that we deploy another system to a problematic site that they were dealing with at the time (a borehole well with high concentrations of CO_2), (*location C*). Stakeholder involvement is time-consuming in building up a working relationship After calibrating our sensors we were then ready to deploy 3 landfill systems into the physical environment. From our experience with previous developmental models [26,29,30], we were confident that the systems were sufficiently robust to withstand long term deployments in the environment.

Initially we deployed one landfill unit to *location A* on the 28th May 2009, which sampled twice a day (11 AM + 11 PM) until the 8th October 2009. From the 13th August 2009 we then deployed a second landfill unit to *location B* to concurrently sample twice a day (11 AM + 11 PM) (Our field-trial deployments are listed in Table 1, and will be explained in more detail in Section 4.3). After these two trials, we recalled the units to our research labs to carry out some maintenance work after finding that insects had breached the system during a regular battery change. At this point, the systems were thoroughly examined, cleaned and the PCB boards were protected by a layer of spray silicone (Electrolube ERDCA200H). Between November and December we redeployed one landfill unit to *location A*. After further maintenance, from early March 2010 until the present time we have been sampling at 4 times every day (12 AM, 6 AM, 12 PM, 6 PM) in both *location A* and *location B*. From May 24th we added a third location, *location C*, meaning 3 locations were being sampled 4 times per day at 12 AM, 6 AM, 12 PM, and 6 PM.

Deployment Data Processing

As described earlier, each monitoring cycle consisted of a 3 minutes *baseline* (60 samples taken at 0.33 Hz), a 3 minutes *sample* (60 samples taken at 0.33 Hz), and a 3 minutes *purge* (60 samples taken at 0.33 Hz), with a statistical representation of each of the 3 stages being sent back to our central base-station via GSM. Meanwhile the fully recorded dataset was stored in on-board flash memory in the landfill system, and then downloaded at a later date for further analysis. We then had the ability to take this data back to our research labs (downloaded during battery changes) to carry out post-event analysis and determine what the optimum sampling rate should be for future deployments of our landfill systems. The objective of this exercise was to improve battery lifetime, and reduce transmission and processing costs.

To carry out our computational analysis we considered data from *location A* between 29th July 2009 and 9th October 2009. This equated to 143 9-minute monitoring cycles (baseline + sample + purge) taken over this period of time. In this exercise, we considered the raw sensor data (recorded every 3 seconds) which equated to 25,740 readings. A software processing algorithm was then used to go through 95 scenarios on all 143 × 60 baseline/sample/purge readings, thus representing optimization investigations on 2,445,300 simulated data readings. In the results section we will report on our findings as to the optimal sampling rate.

As noted in Table 1 there was a period of time in our *location C* deployment where there was a 9 minutes sampling period to/from the borehole well (with no baseline extraction beforehand or purge extraction afterwards). These 93 instances (16,020 raw data samples) allowed us the opportunity to investigate whether the nature of the sample stage remains similar without baseline and purge stages.

This could potentially allow for many savings in terms of shorter sampling times (power, memory and communications loads) and being able to disregard valves (component cost, complexity).

Table 1. Field deployment data gathered over a 16 month period. For *location C* a 9 minutes *sample only* approach was taken, as opposed to *baseline* + *sample* + *purge.*

Location	Start Time	End Time	Num Data Readings	Sampling Rate	CO_2 Avg %	CH_4 Avg %	CO_2 Limit Exceeded	CH_4 Limit Exceeded	CO_2 & CH_4 Exceeded
A	28-May-09	08-Oct-09	255	2× /day	7.71	13.70	185	133	133
B	13-Aug-09	08-Oct-09	113	2× /day	3.23	0.13	77	0	0
A	20-Nov-09	28-Dec-09	77	2× /day	5.99	0.85	58	10	10
C*	03-Mar-10	07-Sep-10	764	4× / day	3.78	0.09	738	0	0
A	10-Mar-10	07-Sep-10	768	4× / day	1.52	0.30	243	47	47
B	24-May-10	07-Sep-10	446	4× / day	1.48	0.02	189	0	0
SUM			2,403				1,490	190	190

RESULTS AND DISCUSSION

Chemical Sensors

The calibration routine of the chemical sensors was described earlier (Section 3.1). At each of the 10 calibration steps we extracted the data when each sensor arrived at a steady state response. Once this was achieved, we noted the respective gas concentration levels from our reference instrument (GA2000 Plus). A 10 point calibration plot for each IR sensor, CO2 and CH4, is presented in Figures 7 and 8, respectively. Excellent correlation between the reference system and landfill sensors was obtained for CO_2 (R2 = 0.99818, n = 10), and for CH4 (R2 = 0.99994, n = 10). It is clear that the system's detection performance is on par with the currently used reference instrument when detecting these two chemical targets. Finally, the linear correlation equations generated by these calibration plots were used as inline conversions from reported ADC measurements by the remote systems for presenting real concentration values online.

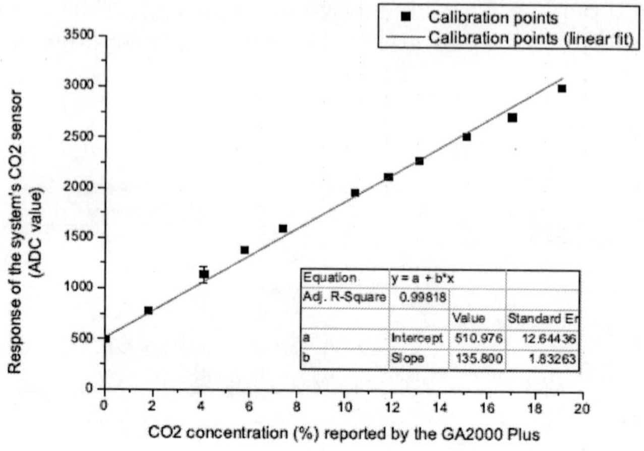

Figure 7. Calibration of the system's CO_2 infrared gas sensor. Points represent the average of the steady state response over circa 2 minutes. Error bars (present but difficult to see due to the high sensor accuracy) represent the standard deviation.

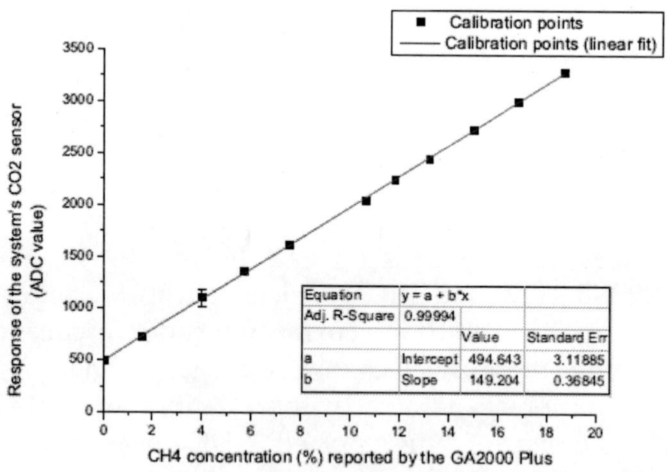

Figure 8. Calibration of the system's CH_4 infrared gas sensor. Points represent the average of the steady state response over circa 2 minutes. Error bars (present but difficult to see due to the high sensor accuracy) represent the standard deviation.

Power Consumption Analysis

Figure 9 shows the current consumption analysis during an aforementioned monitoring and reporting routine (Section 3.2). One can see identifiable trends relating to the four procedures (baseline, sample, purge and communications), and also the times when the extraction and exhaust valves were toggled.

The average current consumption during this time was found to be circa 230.1 mA over a duration of 9 minutes and 45 seconds (585 seconds in total). By the same analysis method, when in its inactive state, the multimeter reported an average current draw of 6.13 mA for 42,615 seconds before the next sampling routine. This was calculated to be an average continuous current draw, from the 12 V source, of 9.16 mA. Assuming an ideal power source with these characteristics, the system (with its present power source and sampling routine) can autonomously monitor landfill gas concentrations for an estimated 4.5 weeks which was a sufficient deployment time (without requiring a battery change) to explore this proof of principle study.

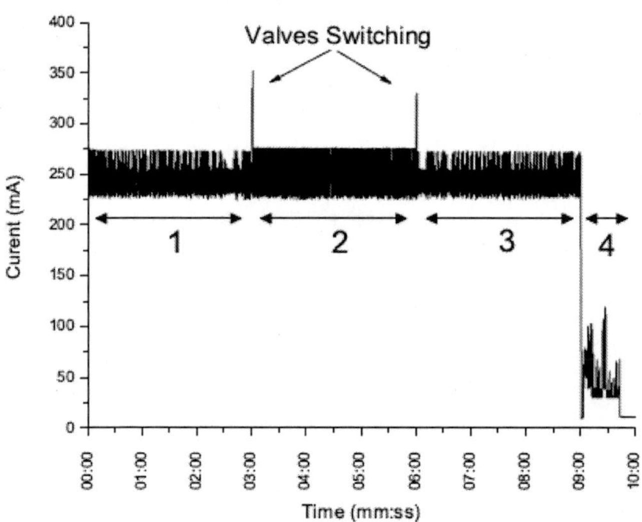

Figure 9. Current analysis of the landfill system during a full monitoring routine (1) baseline procedure, (2) sampling procedure, (3) purge procedure, (4) communications and storage procedure.

Deployment Data

Table 1 summarises the collection of data over a 16 month period as described in Section 3.3. Overall, we can observe that 2,403 samples were sent to the central server, in which the CO_2 limit was exceeded in 1,490 (62%) samples! The CH_4 limit was exceeded on 190 (8%) occasions, while both CO_2 and CH_4 were exceeded together on 190 (8%) occasions. To consider an individual deployment, we illustrate all the sampled readings from the *location* C field deployment in Figure 10. In this case, the CO_2 component exceeded the recommended limit [24] in 96.6% of the samples, while CH_4 never exceeded the recommended limit [24] *i.e.*, 0% of the time. The average CO_2 value recorded was 3.78%, which is 2.52 times above the regulatory limit of 1.5% v/v. The average CH_4 value was 0.01% which is within the regulation limit of 1% v/v. It is worthwhile to note that, CO_2 levels in soil/sub soil layer can naturally exceed 1.5% due to a number of external processes e.g., aerobic degradation of organic matter in soil, dissolution of CO_2 from groundwater high in carbonate, microbial methane oxidation. Thus CO_2 levels above 1.5% do not necessarily indicate landfill gas migration, however our methodology follows well established procedures and pre-existing monitoring routines by the EPA. An ideal solution would be to investigate typical background levels in the area being monitored, which are unaffected by the landfill. Furthermore, the levels quoted are limits for air; the borehole levels tell us the concentrations of these gases migrating within the landfill site that could be released into the air, and could be inherently dangerous if left uncontrolled [31].

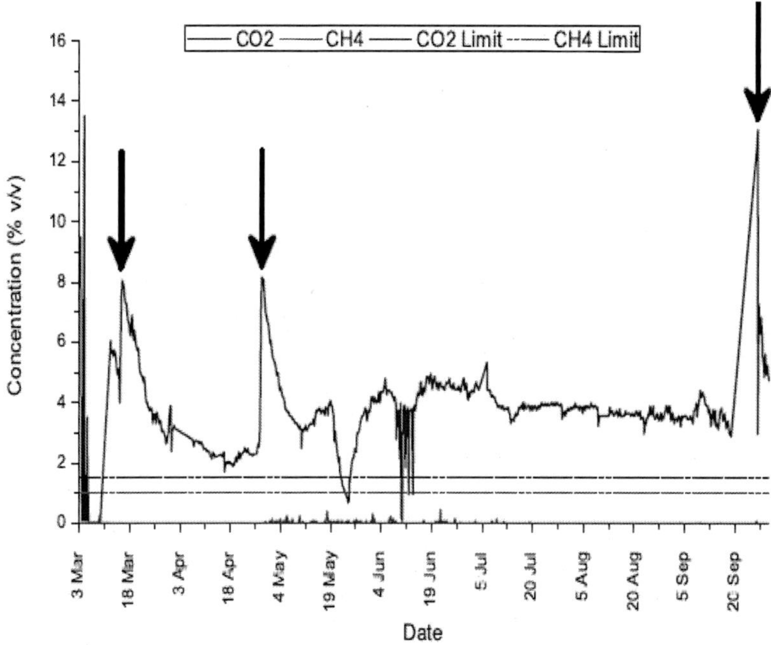

Figure 10. CO_2 and CH_4 readings from a 7 month field deployment at *location C* between March 2010 and October 2010. Note that CO_2 exceeds the recommended limit 96.6% of the time, while CH_4 never exceeds the recommended limit. The arrows on the graph illustrate significant CO_2 events that were recorded around the 17th of March, 28th April, and 25th of September. There were no CH_4 events.

Even considering this, observing the 7 month trend of sampled data, significant CO_2 events were recorded around the 17th of March, 28th April, and 25th of September. Greenhouse gas emissions from landfills are inherently dynamic (especially during their initial phase) and events such as these can be attributed to a number of factors including: borehole proximity to the landfill, time of year, seal of the borehole well cap, water table, head-space, sample depth as well as human activities and extraction system failures/blockages.

It is difficult to pinpoint the exact cause(s) of these events at this early stage in our investigations, but clearly, the availability of this type of information will open the way to gaining a fuller understanding of the dynamics of greenhouse gas generation in, and therefore more effective management of, landfill sites. This only strengthens the need for this type of real time monitoring technology. Finally, one important issue that arises from this data series is: how many of these events are missed by the current manual monitoring frequency of once per month. The next section explores this question.

Human Operator Error Simulation

Considering Figure 11, in which we simulate a human operator taking a reading on a particular first day of each given month, it can be seen that many dynamic events would not be noticed particularly for the middle of March-2010, the end of April-2010, and the middle of September. For example if a human operator noted a reading on the firstMonday of every month (at 12 noon), then there would be an average error of 7% (4.05% for CO_2 *vs.* actual average reading of 3.78% from all sampled data points over 7 months). The first Tuesdays in our dataset would have yielded an average error of 11%, Wednesdays an error of 35%, Thursdays 33%, and Fridays an average error 0%. From these 5 scenarios alone (plus a visual inspection of Figure 11), it can be seen that there is a wide degree of error in selecting a manual rota for human operators to monitor overall landfill emissions.

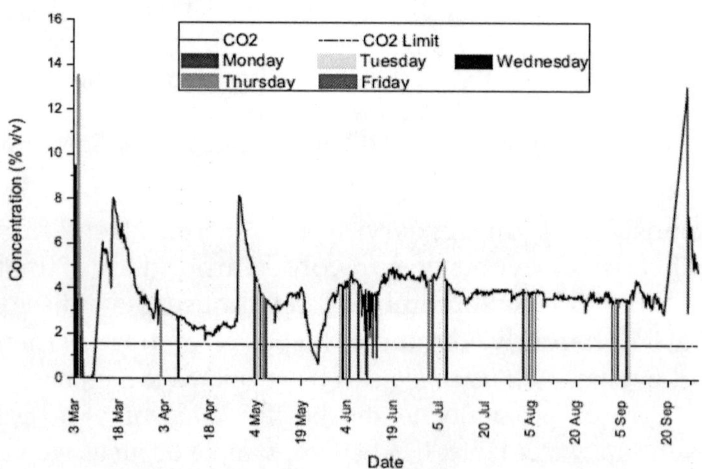

Figure 11. CO_2 readings over a 7 month field deployment at *location* C betweenMarch 2010 and October 2010. Note that a simulated human operator sampling the landfill emissions on a particular first day or each month

would miss a lot of events of interest e.g., the middle of March-2010, the end of April-2010, and the middle of September. The average error across each of the 5 days (Mon-Fri) would have been 17% in our field deployment at *location* C.

Sampling Procedure Analysis

The sampling method used in this study has been developed on the back of detailed previous investigations into how best to sample the geochemical gas composition levels so that a representative and an accurate analysis is obtained [24,26,32]. These have shown that the sensors and sampling routine are not affected by other parameters such as flow rate or pressure, but by the sensing targets within the borehole. In addition, although this study focused on our national (Irish) acceptable gas emission levels of 1% for methane and 1.5% for carbon dioxide, *i.e.*, to reflect deployments locally, it should be noted that these acceptable levels are similar in many countries such as the United Kingdom [33], Canada [34], India [35], *etc.* with a similar recommended sampling frequency of once per month. It is also interesting to note that all have limited the methane level at 25% of its lower explosive limit *i.e.*, between 1% and 1.25% v/v.

To carry out our computational analysis we considered data from *location A* between 29th July 2009 and 9th October 2009. As described, our system firstly carried out a 3 minutes *baseline* stage, followed by an actual 3 minutes *sample* stage, and finally a 3 minutes *purge* stage (see Figure 12). We now discuss the CO_2 and CH_4 profiles associated with each of those stages. Note: all data presented here is from the full dataset downloaded from the field-deployed landfill units.

Baseline Stage

The average *baseline* profile (across the entire 143 recorded readings) is illustrated on the left hand side of Figure 12. Throughout our field deployment, the transmitted CO2 and CH_4 *baseline* readings were calculated by taking the last 11 readings (33 seconds) of the full dataset (*i.e.*, after the sensors had time to warm up and just before the sampling stage) and calculating the average, however on post-analysis inspection of the baseline profile in Figure 12 it appeared we could get near that average by taking fewer samples. We then ran a software simulation program which went through 95 variations, on all 143 × 60 *baseline* readings. Our finding was that after 20 readings, and taking the average of the last 5 samples we then achieve a very low average reading error of 0.17% for CO_2 and 0.62% for CH_4, with an individual outlier worst case of 2.51% (CO2) and 2.57% (CH4) error in ADC reading, compared to the field-deployment implementation. This means that we have identified the point at which the sensors achieve a steady state response after a necessary and unavoidable warm up period. Overall, this represents a battery saving of over 60% for the *baseline* sampling stage alone.

Sample Stage

Next, the average *sample* profile (across the entire 143 recorded readings) is illustrated in the centre of Figure 12, for the chemical and physical sensors, respectively. Throughout our field deployments, the CO_2 and CH4 composition levels were calculated by taking the maximum CO_2, CH_4, Humidity & Temperature readings (60 3-second samples taken in total). However, on inspection of the sample stage in Figure 12 it appeared that we could get those maximum values through taking less samples by visual inspection of the trends alone. We then ran a software simulation program which went through 95 variations, on all 143 × 60 *sample* readings. Our finding was that after 30 readings, and taking the average of the maximum sampled values, an average error of 0.72% (CO_2), 0.34% (CH_4), 0.07%

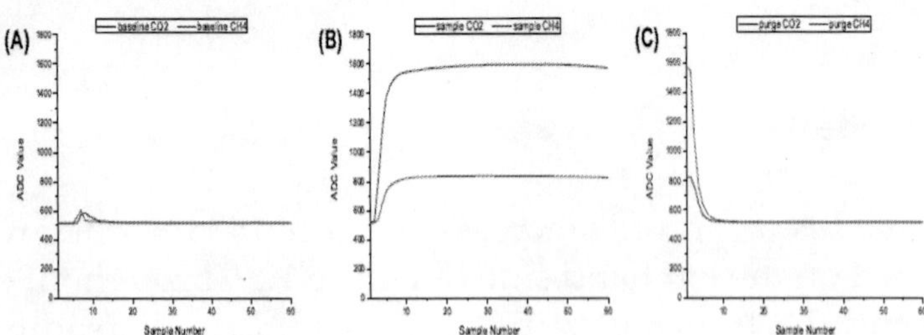

Figure 12. Profile of a typical 9 minutes *baseline* (A), *sample* (B), & *purge (C)* sampling

Stage, which is comprised of 180 CO_2 & CH_4 samples recorded every 3 seconds. This occurs in the order of 60× *baseline*, 60× *sample*, and 60× *purge* samples. Initially all 180 items were sampled, however after a close analysis of 10+ weeks of data, we have been able to minimise the length of this sampling procedure. This has a positive effect on battery power consumption.

Purge Stage

Lastly, the average *purge* profile (across the entire 143 recorded readings) is illustrated on the right hand side of Figure 12. Throughout our field deployment, CO_2 and CH_4 levels were calculated by taking the minimum recorded readings (60 3-second samples taken in total), however on post-analysis inspection of the purge stage in Figure 12 it appeared we could get such a representative value by taking less samples. We then ran a software simulation program which went through 95 variations, on all 143 × 60 *purge* readings. Our finding was that after 20 readings, and taking the minimumrecorded value we then get an average reading error of only 0.28%for CO_2 and 0.12%for CH_4, with an individual outlier worst case of 2.74% (CO_2) and 0.78% (CH_4) error in ADC reading, compared to the field-deployment implementation. This represents a battery saving of over 60% for the *purge* sampling stage.

So in summary, if we consider using a system as follows: 1 minute baseline; 1.5 minutes sample; and 1 minute purge, we would be within an average sample stage error of 0.72% (CO_2), 0.34% (CH_4), 0.07% (Humidity) and 0.007% (Temperature), with a worst case of 2.88% (CO2) in ADC readings. This would mean that our total field trial samples could be reduced from 10,010, as opposed to 25,740 which would represent a potential extension of battery life by 2.57 times.

Removal of Baseline & Purge Stages

As noted in Table 1 there was a period of time in our *location C* deployment where there was a 9 minutes monitoring period (with no baseline extraction beforehand or purge extraction afterwards). Comparing the very similar signatures of the sample stage in Figures 12 (*baseline* + *sample* + *purge*) and 13 (*sample only*), there is an indication that only the sample stage is needed. Carrying out a software optimisation of a *sample only* system, we found that just 30 (× 3 seconds) readings are required. The first 10 readings (*i.e.*, 30 seconds) are required for the sensors to heat up to a steady state and also for the pump to flush the sampling chamber in order to eliminate any 'memory effects' that may be present from the previous sampling stage.

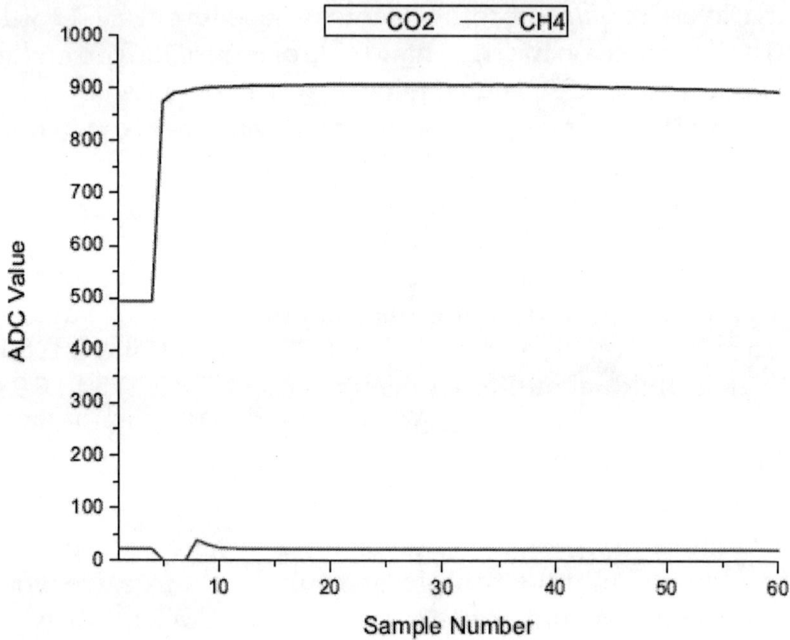

Figure 13. Profile of our *"sample only"* system, note the same signature as Figure 12, which possibly indicates that only the *sample* stage is needed to measure the air emissions at a given landfill site.

Sampling Frequency

The initial use of this platformwas to provide remote access to accurate data for enforcement purposes (EPA); four samples per day is a great improvement on one sample per month (EPA's current procedure).

It is also a balance between sampling rate and power demand we anticipate that the incorporation of the solar panel will enable the sampling frequency to be increased towards the levels required for more effective modelling of the site and optimisation of operation. In that respect, there may be more value from the monitoring of other targets such as pressure at higher frequencies as elevated levels of gases are often associated with blockages in the extraction system.

Lessons Learned

The evaluation procedure explored previously is not sufficiently comprehensive at this stage to make a definite case to drop the baseline and purge stages. The most effective way to do this would be to run the 2 systems in parallel and verify that both are analysing the same landfill gas. However, from all the data and experience available to us, we feel that a *sample only* system is the best approach to take in future deployments. We recommend a sampling stage as follows (at least 4 times per day):

1. Allow 30 seconds for sensors to warm up (no sampling required)

2. Sample every 3 seconds for the next 90 seconds (CO_2, CH_4, Humidity, Temperature)

3. Stop if 5 consecutive readings report the same values

4. Record the maximum reported values from step 2 for CO_2, CH_4, Temperature, and Humidity Such a system, compared to that deployed in our field trials (9 minutes samples with *baseline, sample,* and *purge* stages), would offer a number of advantages:

• No valves required, so less mechanical complexity and cost, and increased reliability and battery life.

• Total sample time reduced from 9 minutes to 1 m 45 seconds A saving of 80% in active power consumption, and potentially increasing battery field lifetime from 6 weeks to approximately 30 weeks; this figure does not factor in power savings achieved through not having to actuate valves.

• Reduced manufacturing costs and increased battery life and reliability, as switching valves are no longer required.

Future Work

Communications

At present we have achieved remote data retrieval through compiling a statistical representation of the data and by transmitting to the base station by means of SMS. The SMS text structure was formed so that it could be readily interpreted by a human observer. Although this option was beneficial at early stages of the project, we have since progressed to a complete data orientated formulation where human observations are now at the visualisation end (see Figure 6). As a result, one can retrieve a richer sensor data-set by introducing encoding schemes (such as Huffman or Arithmetic) to compress the data and ultimately retrieve more information per transmission. This will reduce the transmission frequency resulting in lower cost and power use. Furthermore, this can potentially result in retrieval of fully recorded datasets via SMS without the user physically being present.

Alternatively when we deploy multiple units on a site an additional communications layer should be considered, whereby the units will be equipped with low power, short range radio transceivers (such as Zigbee). Then each sensor would report all its findings to a central communications gateway over a star/mesh/bus wireless sensor network as outlined by a recent survey [36]. We foresee that the communications gateway will relay all data to the base-station via GPRS/3G on sites without any local access points. On the other hand, many active sites have a nearby workplace with internet capability where one can potentially take advantage of new generation technologies such as WiMAX [37]. This strategy will give a new layer of scalability to the sensing structure and allow many other plug and play sensing nodes to be added (such as more sensed locations, wider range of gases e.g., H_2S, gas pressure monitoring *etc.*).

Ultimately, the chosen communications method will depend on the layout of the site and also on the number of nodes needed. At this stage in our deployments, SMS communications has been found to be sufficient to explore this application principle.

Adaptive Sampling

Borehole measurements typically involve monitoring the sensor output until a steady state signal is achieved for a representative gas sample. Many factors have been found to affect this and have been listed earlier in Section 4.3. A subset of these has already been explored in a previous study [26] and to accommodate these factors, we have chosen to sample for a period of 3 minutes. However, there are disadvantages to this approach with the primary drawback being unnecessary power usage on wells that generate a steady state signal relatively quickly. This ultimately reduces the lifetime of the device in the field.

Although we have addressed this issue earlier (Section 4.5), we foresee a further extension of an overall adaptive sampling technique including a varied sampling cycle frequency, *i.e.*, a fixed frequency of 2 times per day may be too high for some sites and too low for others. We foresee the application of computer classification techniques to adaptively select the optimum sampling frequency, based on previous reported gas concentrations and battery capacity at the time.

Energy Reduction/Harvesting

It has been determined earlier (Section 4.2) that the system can function autonomously for circa 4.5 weeks using the current power source. Although this was acceptable for our purposes, it is desirable to maximise the functional lifetime of the system (where possible) in the environment. The principle reason for this is that the cost of maintenance al one (for battery changes and the human resources required to change them) can be substantial, especially with multiple sites. A future option to explore is the harvesting of energy through using devices such as solar panels in the first instance. This is a crucial limiting factor, as scalability depends on sensors being able tomeet their operational energy requirements from integrated energy generation capabilities. Our most recent efforts have made significant headway with the integration of a solar panel and anticipate that we will be deploying systems in the field in the near future for evaluation. Laboratory data suggests that standard panels will be able to meet the power demand of the platform and dramatically extend its operational lifetime.

Integration of Other Sensors It should be noted that CO_2 and CH_4 were the primary sensing targets as identified by the EPA; they are also accessible via IR sensors that are very reliable and as we have demonstrated, suitable for long term autonomous deployment (greater than 1 year) with platforms that are relatively low cost.

Inclusion of additional targets means complicating the sensing platform, potentially driving up the cost significantly, and reducing the capacity to function autonomously, which is directly against the goal of the project. So while we appreciate that additional targets are important, we had to strike a balance between long term viability and number of targets.

Conceptually, our setup can accommodate many more sensor types (with minimum alterations to the gateway platform) and also many more gateway platforms. Also, individual components such as the communications module can be swapped out very easily to accommodate other standards.

Recently, we have equipped a landfill unit with other chemical gas sensors to monitor levels of ammonia (NH_3) and hydrogen sulphide (H_2S) by introducing simple signal conditioning circuitry to the system, and we are ready to interface the system to the portal page once a suitable location has been found. In addition, we have already expanded the sensor server and web visualisation interface to include real-time monitoring of phosphate (PO_4^{3-}) in river waters, carbon monoxide (CO) in car parks, domestic

carbon emissions and pressure of extraction systems on landfill sites [38].

FUTURE DEPLOYMENTS

This study has provided the means to fully equip landfill sites with multiple analyser platforms to detect landfill gas migration and/or surface emissions. In the first instance, our aim is to increase production of the systems and launch a case study on a young, active landfill site. This will allow us to harvest data across an entire site and over a long period of time with an end goal to apply software modelling techniques. Ultimately this may provide a real enhancement of landfill management capabilities, e.g., through the development of early warning systems. Furthermore, with such a deployment, we may be able to identify many more aspects of landfill dynamics such as seasonal effects, tears in liners during installation, blocked pipes *etc*.

CONCLUSIONS

We have successfully realised and validated a platform for real time monitoring of landfill subsurface migration gases. Our system incorporates sensing of carbon dioxide and methane emissions at landfill sites, GSM communications to a "cloud database" and an on-line visualisation element to deliver near real time data to users in an easy to interpret format. This system has been successfully deployed in field-tests over a 16 month period, with 3 separate devices running concurrently across multiple locations towards the end of the trial generating 2,445,300 *in-situ* measurements of gas concentrations during this time.

Through post analysis of the data gathered over a 16 month period, we have identified further

improvements that we can make to the system to reduce the cost and power consumption. The advantages of our remote monitoring system is that it consistently gathers data at a much higher granularity than the current manual sampling regime thereby reducing the risk of missing events, in addition to reducing the possibilities for human operator error. From experience "at the coalface" through long-term field trials, and through involvement with relevant environmental and industrial stakeholders, our system has evolved to the stage where we are confident that it is now possible for authorities to complement their manual monitoring with a much more rich stream of data.

Finally, one conclusion of particular interest is that we have shown that this approach, with much

higher sampling frequency, opens up a rich source of new environmental information about the dynamics of landfill gas generation and migration, that can ultimately lead to a better understanding of the processes underlying "events" (rapid increases in emissions), and therefore, more effective management of these facilities.

In addition, recent events have shown that catastrophes can and do occur at landfill sites, *i.e.*, a massive underground fire erupting at a nearby 50-acre site in Co Kildare, Ireland. This has had an overall negative impact in many ways such as health risks to thousands of nearby residence through pollution of the surrounding air, evacuations and it has been estimated that it will cost more than 30 million euros to recover from this disaster [31].

It is clear that we must rethink our approach to effectively and efficiently managing our domestic waste responsibilities via land filling, to one where we start to utilise the extensive capabilities of the sensor research community. The approach proposed in this study aims at providing a means to that end.

ACKNOWLEDGMENTS

This material is based upon work supported by Science Foundation Ireland under Grant No. 07/CE/I1147. Author Aiden R. Doherty is supported by the Irish Health Research Board under Grand No. MCPD/2010/12. Our environmental data sensor portal is available at http://clarityapp.ucd.ie/~sensorportal/.

REFERENCES

1. Solomon, S.; Qin, D.; Manning, M.; Chen, Z.; Marquis, M.; Averyt, K.B.; Tignor, M.; Miller, H.L. *Climate Change 2007: The Physical Science Basis. Contribution of Working Group I to the Fourth Assessment Report of the Intergovernmental Panel on Climate Change*; Cambridge University Press: Cambridge, UK; New York, NY, USA, 2007.
2. Austin, B. The Involvement of Pollution with Fish Health. In *Multiple Stressors: A Challenge for the Future*; Mothersill, C., Mosse, I., Seymour, C., Eds.; Springer: Dordrecht, The Netherlands, 2007; pp. 13-30.
3. Yasuda, M.; Hanagiri, T.; Shigematsu, Y.; Onitsuka, T.; Kuroda, K.; Baba, T.; Mizukami, M.; Ichiki, Y.; Uramoto, H.; Takenoyama, M.; Yasumoto, K. Identification of a Tumour Associated Antigen in Lung Cancer Patients with Asbestos Exposure. *Anticanc. Res.* **2010**, *30*, 2631-2639.
4. Kazakov, V.S.; Demidchik, E.P.; Astakhova, L.N. Thyroid cancer after Chernobyl. *Nature* **1992**, *359*, 21-22.
5. Khurana, V.G.; Hardell, L.; Everaert, J.; Bortkiewicz, A.; Carlberg, M.; Ahonen, M. Epidemiological evidence for a health risk from

mobile phone base stations. *Int. J. Occup. Environ. Health* **2010,** *16,* 263-267.

6. The European Parliament. The EU Water Framework Directive. 2000. Available online: http://ec.europa.eu/environment/water/water-framework (accessed on 20 June 2011).
7. The United Nations. Kyoto Protocol. 1997. Available online: http://www.unfccc.int (accessed on 20 June 2011).
8. Department for Environment Food and Rural Affairs. Implementing the Climate Change Act (UK). 2008. Available online: ttp://www.defra.gov.uk/environment/climate/legislation/ (accessed on 20 June 2011).
9. California Environmental Protection Agency. Assembly Bill 32: Global Warming Solutions Act. 2006. Available online: http://www.arb.ca.gov/cc/ab32/ab32.htm (accessed on 20 June 2011).
10. Byrne, R.; Diamond, D. Chemo/bio-sensor networks. *Nat. Mater.* **2006**, *5*, 421-424.
11. Diamond, D. Internet-scale sensing. *Anal. Chem.* **2004**, *76*, 278-286.
12. Doherty, A.R.; Smeaton, A.F. Automatically Segmenting Lifelog Data Into Events. In *Proceedings of WIAMIS: 9th International Workshop on Image Analysis for Multimedia Interactive Services*, Klagenfurt, Austria, 7–9 May 2008.
13. Lou, X.; Nair, J. The impact of landfilling and composting on greenhouse gas emissions—A review. *Bioresource Technol.* **2009**, *100*, 3792-3798.
14. Pipatti, R.; Wihersaari, M. Cost-effectiveness of alternative strategies in mitigating the greenhouse impact of waste management in three communities of different size. *Mitigation Adaptation Strategies Global Change* **1997**, *2*, 337-358.
15. . Pipatti, R.; Vieira, S.M.M. Waste. In *IPCC Guidelines for National Greenhouse Gas Inventories*; Eggleston, S., Buendia, L., Miwa, K., Ngara, T., Tanabe, K., Eds.; The Intergovernmental Panel on Climate Change: Washington, DC, USA, 2006.
16. Crouch, E.A.C.; Green, L.C.; Zemba, S.G. Estimation of Health Risk from Landfill Gas Emissions. In *Proceedings of the GRCDA 13th Annual International Landfill Gas Symposium,* The Association of Soil Waste Management Professionals: Lincolnshire, IL, USA, 27–29 March 1990.
17. The Environmental Protection Agency. Landfill Manual—Site Design. 2000. Available online: http://www.epa.ie/downloads/advice/licensee/name,14447,en.html (accessed on 20 June 2011).
18. Aitchison, E. Methane generation from UK landfill sites and its use as an energy resource. *Energy Conver. Manag.* **1996**, *37*, 1111-1116.
19. Brjesson, G.; Sundh, I.; Svensson, B. Microbial oxidation of CH4 at different temperatures in landfill cover soils. *FEMS Microbiol. Ecol.* **2004**, *48*, 305-312.
20. Vrijheid, M.; Armstrong, B.; Dolk, H. Hazard scoring of landfill sites in relation to risk of congenital anomalies near hazardous waste landfill sites in Europe. *Am. J. Epidemiol.* **1998**, *147*, 297.

21. Martuzzi, M.; Mitis, F.; Forastiere, F. Inequalities, inequities, environmental justice in waste management and health. *Eur. J. Public Health* **2010**, *20*, 21-26.
22. Vrijheid, M.; Dolk, H.; Armstrong, B.; Abramsky, L.; Bianchi, F.; Fazarinc, I.; Garne, E.; Ide, R.; Nelen, V.; Robert, E.; Scott, J.; Stone, D.; Tenconi, R. Chromosomal congenital anomalies and residence near hazardous waste landfill sites. *Lancet* **2002**, *359*, 320-322.
23. Dolk, H.; Vrijheid,M.; Armstrong, B.; Abramsky, L.; Bianchi, F.; Garne, E.; Nelen, V.; Robert, E.; Scott, J.; Stone, D.; Tenconi, R. Risk of congenital anomalies near hazardous-waste landfill sites in Europe: the EUROHAZCON study. *Lancet* **1998**, *352*, 423-427.
24. The Environmental Protection Agency. Landfill Manual—Guidance note of Landfill Monitoring. 2003. Available online: http://www.epa.ie/downloads/advice/licensee/name,14437,en.html (accessed on 20 June 2011).
25. The European Council. The Landfill Directive. 1999. Available online: http://ec.europa.eu/environment/waste/landfill index.htm (accessed on 20 June 2011).
26. Kiernan, B.; Beirne, S.; Fay, C.; Diamond, D. Measurement of Representative Landfill Gas Migration Samples at Landfill Perimeters: A Case Study. In *Proceedings of The 24th International Conference on Solid Waste Technology and Management*, Philadelphia, PA, USA, 15–18 March 2009.
27. Dynament Ltd. Carbon Dioxide InfraRed Gas sensor Product Specifications. 2011. Available online: http://www.dynament.com/infrared-sensors/carbon-dioxide/premier-certified-carbondioxide-sensors.htm (accessed on 20 June 2011).
28. Dynament Ltd. Methane InfraRed Gas sensor Product Specifications. 2011. Available online: http://www.dynament.com/infrared-sensors/methane/premier-certified-methane-sensors.htm (accessed on 20 June 2011).
29. Kiernan, B.; Fay, C.; Beirne, S.; Diamond, D. Development of an Autonomous Greenhouse Gas Monitoring System. In *Proceedings of World Congress on Science, Engineering and Technology*, Venice, Italy, 29–31 October 2008.
30. Kiernan, B.; Beirne, S.; Fay, C.; Diamond, D. Landfill Gas Monitoring at Borehole Wells using an Autonomous Environmental Monitoring System. In *Proceedings of The 5th International Conference on Climate Change and GlobalWarming*, Frankfurt, Germany, 12–14 September 2008.
31. Hogan, T. Fears over fire that can't be put out. 2011 Available online: http://www.independent.ie/national-news/fears-over-fire-that-cant-be-put-out-2511027.html (accessed on 20 June 2011).
32. The Environmental Protection Agency. Monitoring of Gas Emissions at Landfill Sites Using Autonomous Gas Sensors. 2010. Available online: http://www.epa.ie/downloads/pubs/research/tech/name,28454,en.html (accessed on 20 June 2011).

33. Rosevear, A.; Deed, C.; Gronow, J.; Keenlyside, J.; Smith, R.; Braithwaite, P. Guidance on Monitoring Landfill Gas Surface Emissions. 2010. Available online: http://publications.environment-agency.gov.uk/PDF/GEHO0311BTOL-E-E.pdf (accessed on 20 June 2011).

34. Ministry of Environment Guidelines for Environmental Monitoring at Municipal Solid Waste Landfills. Available online: http://www.env.gov.bc.ca/epd/epdpa/mpp/gfemamsw.html (accessed on 20 June 2011).

35. Ministry of Environment and Forests. Municipal SolidWastes (Management and Handling) Rules. 1999. Available online: http://http://www.moef.nic.in/ (accessed on 20 June 2011).

36. Yick, J.; Mukherjee, B.; Ghosal, D. Wireless sensor network survey. *Comput. Netw.* **2008**, *52*, 2292-2330.

37. Nissila, T.; Pentikousis, K.; Harjula, I.; Huusko, J.; Rautio, M. WiMAX Backhaul for Environmental Monitoring. In *Proceedings of the 7th International Conference on Mobile and Ubiquitous Multimedia*, Umea, Sweden, 3–5 December 2008.

38. CLARITY: Centre for Sensor Web Technologies. Environmental Sensor Portal Page. 2010. Available online: http://clarityapp.ucd.ie/~sensorportal/ (accessed on 20 June 2011).

Chapter 2

RETRIEVAL OF SURFACE AIR SPECIFIC HUMIDITY OVER THE OCEAN USING AMSR-E MEASUREMENTS

Masahisa Kubota * and Tsutomu Hihara

School of Marine Science and Technology, Tokai University/ Orido, 3-20-1, Shimizu, Shizuoka,

Shizuoka, JAPAN 424-8610; E-Mail: tsutomu@ mercury.oi.u-tokai.ac.jp (T. H.)

* Author to whom correspondence should be addressed; E-Mail: kubota@mercury.oi.u-tokai.ac.jp

(M. K.); Tel.: +81-54-337-0916; Fax: +81-54-338-9983

ABSTRACT

We have developed a new algorithm to estimate the surface air specific humidity over the ocean from AMSR-E data. It should be noted that remarkably reduced random errors of the estimated surface air specific humidity result from using the surface air specific humidity provided by reanalysis data. We validated our new algorithm using independent ship and buoy data. The bias, RMS error, and correlation coefficient of the products obtained using our algorithm for global buoys are 0.38 g/kg, 0.61 g/kg and 0.99, respectively. It should be noted that surface specific humidity having similar accuracy to the reanalysis data near in situ data could be derived from AMSR-E data by the present algorithm.

INTRODUCTION

Latent heat flux (LHF) is considered to be the most important constituent of ocean surface heat flux, which comprises shortwave radiation, longwave radiation, latent heat flux, and sensible heat flux. This is because LHF contributes to the regulation of global climate by transporting a large amount of heat energy in the form of water vapor from low latitudes to mid and high latitudes. In addition, LHF plays an important role in the global hydrological cycle by transporting water not only from the ocean to the atmosphere but also from one location to another. Therefore, it is necessary to monitor the LHF for better understanding of the variability of global climate.

Recently, the global LHF is estimated by using data from several satellites and the following bulk formula:

$$\mathrm{LHF} = \rho_a L_v C_E U(Q_s - Q_a),$$

where ρ_a is the density of air, L_v is the latent heat of vaporization, *U* is the near-surface wind speed relative to the surface current speed, *Qs* and *Qa* are the surface and saturated specific humidity, respectively, and *CE* is the turbulent exchange coefficient for moisture. In order to estimate the amount of LHF, we need to use three kinds of meteorological data such as wind speed, saturated specific humidity, and specific humidity. The saturated specific humidity data can be estimated from the sea surface temperature (SST) data. Several studies have been performed on the development of algorithms for the estimation of specific humidity, e.g.; [1-3]. Most of these algorithms were developed for application to the data of the Defense Meteorological Satellite Program (DMSP)/ Special Sensor Microwave/Imager (SSMI). DMSP/SSMI data are used in many products, e.g.; Japanese Ocean Flux Data Sets with Use of Remote Sensing Observations (J-OFURO) [4], Goddard Satellite-based Surface Turbulent Fluxes (GSSTF2) [5], and Hamburg Ocean Atmosphere Parameters and Fluxes from Satellite Data (HOAPS) [6]. On the other hand, it should be noted that global microwave data can also be obtained using the Advanced Microwave Scanning Radiometer for EOS (AMSR-E). If we can estimate the specific humidity using AMSR-E data, we can reduce the sampling error in the daily mean value of specific humidity because the observation time of AMSR-E (1 a.m. and 1 p.m.) is quite different from those of DMSP/SSMIs (around 6 a.m. and 6 p.m.). However, to date, only [7] has developed an algorithm for estimation of specific humidity using AMSR-E data. They developed the algorithm by assuming that the values of the reanalysis data are the true values. Several previous studies e.g.; [8] and [9] investigated the accuracy of specific humidity data in reanalysis data.

However, it is not clear whether reanalysis data are more accurate than satellite-derived data. Therefore, it is necessary to evaluate the accuracy of the air specific humidity derived from AMSR-E data using [7]'s algorithm, by comparing with ocean observation data. The objective of our study is to develop an algorithm for the estimation of specific humidity using AMSR-E data. We compare our products with other products and validate all these products using buoy data. The data that we have used in this study are described in the next section. Section 3 presents a new algorithm that is developed by using ship observation data, and shows the evaluation of the results derived from the present algorithms using independent ship observation data. Many products, including our products, are evaluated using buoy data in Section 4. Finally the discussion and summary are given in Section 5 and 6, respectively.

DATA

The data obtained by AMSR-E L1B comprise fundamental physical values observed by the instrument, and in the case of AMSR/AMSR-E, they correspond to brightness temperatures. Although the data are not projected onto a map, but stored in the swath form, data location and quality information are also included in AMSR-E L1B data. In this study, we have used the specific humidity data in the reanalysis data of the National Centers for Environmental Prediction (NCEP)/National Center for Atmospheric Research (NCAR) [10] to obtain information about the surface air humidity. We have also used the in situ surface air specific humidity data, which is observed by ships and buoys, of the International Comprehensive Ocean-Atmosphere Data Set (ICOADS) for the construction of the new algorithm. The ICOADS consists of global marine data observed between 1784 and 2007, primarily from ships of opportunity; the data have been collected, edited, and summarized statistically for each month of each year of the period. ICOADS sampling varies regionally ranging from zero to several thousand samples per 2°x2° box per calendar month. Moreover, in many areas there are time dependent changes in the sampling density [11]. After carrying out QC, the surface specific humidity data of ICODAS are converted to their corresponding values at a height of 10 m using COARE 3.0 [12].

We evaluated many products, including the product obtained by our new algorithms, using buoy data. Figure 1 shows the buoy locations. We used the buoy data of the Kuroshio Extension Observatory (KEO) [13], the National Data Buoy Center (NDBC) [14], the Tropical Atmosphere Ocean/Triangle Trans-Ocean Buoy Network (TAO/TRITON), and the Pilot Research Moored Array in the Tropical Atlantic (PIRATA) [15]. We used the data for the year 2004 for analysis, except in the case of the KEO buoy because the KEO has only been carrying out observations since June 2004.

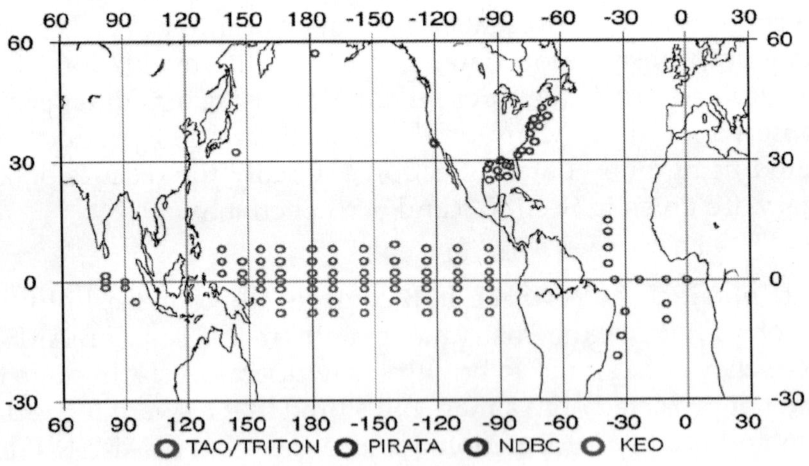

Figure 1. Buoy locations.

DEVELOPMENT OF THE NEW ALGORITHM

We first placed the AMSR-E data and ship observation data in time and space. The temporal and spatial differences between the satellite data and ship observations were limited to less than 30 min and 25 km, respectively. We developed the following regression formula (001) for the collocated satellite and ship observation data. After investigating the most efficient combination of the brightness temperature data observed by the AMSR-E, we finally decided to use all the brightness temperature data in order to obtain an accurate estimation of the surface air specific humidity.

$$Q_a^1 = a_0^1 + a_1^1 T_{6V} + a_2^1 T_{6H} + a_3^1 T_{10V} + a_4^1 T_{10H} + a_5^1 T_{18V} + a_6^1 T_{18H} + a_7^1 T_{22V} + a_8^1 T_{22H} + a_9^1 T_{36V} + a_{10}^1 T_{36H} + a_{11}^1 T_{89V} + a_{12}^1 T_{89H} \quad (1)$$

It should be noted that the vertical profile of humidity affects the accuracy of the estimation of the surface air specific humidity considerably. Since most moisture generally exists near the ground surface, the amount of precipitable water is highly correlated to the surface air specific humidity. Liu [16] performed pioneering work in this field and developed the first algorithm to estimate surface air specific humidity using microwave radiometer data. It should be noted that the difference between in situ and satellite-derived air specific humidity increases if the humidity profile changes because the relation between precipitable water and surface air specific humidity is fixed in this kind of relation. Therefore, we may obtain more accurate specific humidity if we directly use brightness temperature data as the regression formula 001. Reanalysis data may also have different information from that provided by satellite data. Thus, we constructed a second linear regression formula (002) using both the AMSR-E data and Qa (NRA1):

$$Q_a^2 = a_0^2 + a_1^2 T_{6V} + a_2^2 T_{6H} + a_3^2 T_{10V} + a_4^2 T_{10H} + a_5^2 T_{18V} + a_6^2 T_{18H} + a_7^2 T_{22V} + a_8^2 T_{22H} + a_9^2 T_{36V} + a_{10}^2 T_{36H} + a_{11}^2 T_{89V} + a_{12}^2 T_{89H} + a_{13}^2 Q_a(NRA1) \quad (2)$$

where Qa (NRA1) is surface specific humidity provided by the National Centers for Environmental Prediction (NCEP)/National Center for Atmospheric Research reanalysis (NRA1) data. The regression coefficients are shown in Table 1. The value of a2

13 Qa (NRA1) is not so large compared with other terms because the value of brightness temperatures is one digit larger than that of Qa(NRA1)..

However, the effect cannot be neglected. The analysis of variance table is given in Table 2 [17]. The statistics prove that 002 is superior. We validated the products obtained by the two different formulae using the ICOADS data for 2003–2004. These data are independent of the data used in the construction of the formulae because we developed the formulae using the ICOADS data for 2005. The validation was performed for instantaneous data values. Figure 2 shows the scatter plots of the ICOADS observation data for 2003–2004 and the results of the application of the algorithm to AMSR-E data for 2003–2004. For the purpose of comparison, the results for products obtained by applying the algorithms of [1-2,18-20] to AMSR-E data are also given in the same figure. The statistics are provided in Table 3.

Table 1. Regression coefficients.

	a_0	a_1	a_2	a_3	a_4	a_5	a_6
001	-92.775	0.092	-0.067	0.199	-0.181	-0.259	0.310
002	-49.324	-0.003	0.001	0.136	-0.104	-0.118	0.127
	a_7	a_8	a_9	a_{10}	a_{11}	a_{12}	a_{13}
001	1.451	-0.680	-0.908	0.316	0.173	-0.068	
002	0.812	-0.381	-0.524	0.202	0.099	-0.047	0.555

It can be observed that the root mean square (RMS) error remarkably decreases when the algorithm 002 is used. As expected, we can obtain much better results using the new algorithm than the other algorithms. This is expected because the other algorithms were developed not for AMSR-E data but for DMSP/SSMI data. Thus, it can be concluded that we need to develop a different algorithm for each sensor, even if each sensor has channels of similar frequency.

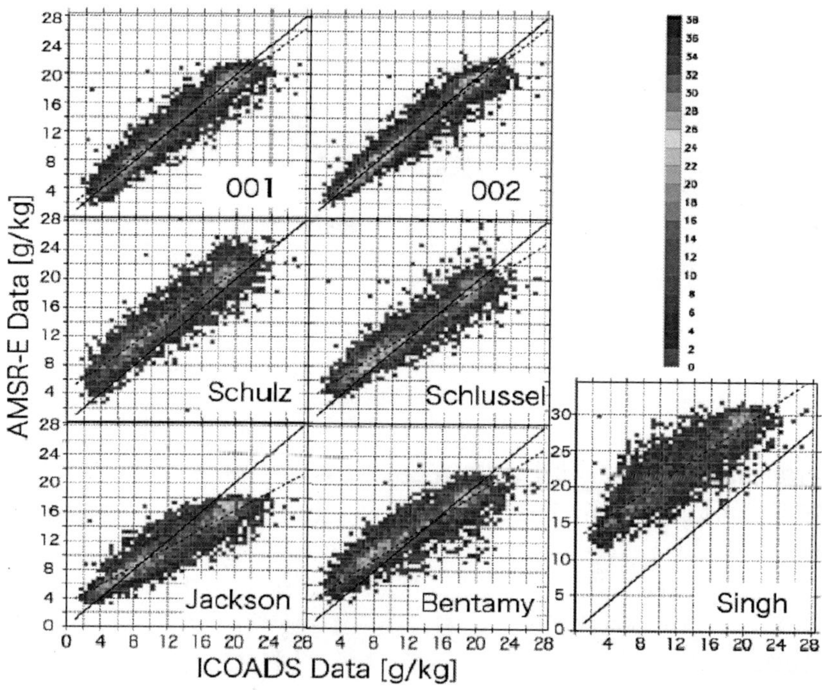

Figure 2. Scatter plots of the surface air specific humidity of the ICOADS data and the data derived from AMSR-E. The color indicates the data number included in a square of 0.5 g/kg x 0.5 g/kg.

Table 2. Analysis of variance (ANOVA) table for simple linear regression of formulae 001 and 002. The column headings df, SS, and MS stand for degree of freedom, sum of square, and mean square, respectively [17].

001	**df**	**SS**	**MS**	**F (=MSR/MSE)**
Total	5,626	170,096		
Regression	12	155,117	12,926 (MSR)	4,845
Residual	5,614	14,979	2.67(MSE)	
002	**Df**	**SS**	**MS**	**F**
Total	5,626	170,096		
Regression	13	158,229	12,171 (MSR)	5,946
Residual	5,613	11,450	2.05(MSE)	

Table 3. Statistics for instantaneous ICOADS specific humidity data and data derived using many different algorithms.

	001	002	Schulz	
BIAS [g/kg]	-0.01	-0.03	2.32	
RMSE [g/kg]	1.75	1.57	2.17	
CORRELATION	0.946	0.957	0.916	
	Schlussel	Bentamy	Singh	Jackson
BIAS [g/kg]	0.45	0.89	10.12	-1.25
RMSE [g/kg]	2.17	2.46	2.65	2.48
CORRELATION	0.919	0.903	0.872	0.912

Comparison with global buoy data

We evaluated various daily products including the present products using buoy data. The statistics are given in Table 4. J-OFURO 2 [21] and HOAPS3 are satellite products, which utilize the data of DMSP/SSMI. We estimated the specific humidity by applying the algorithms of [7] and [18] to AMSR-E data. NRA1, NCEP/Department of Energy reanalysis (NRA2) [22], and Japanese Re-Analysis 25 years (JRA25) [23] data are all types of reanalysis data. Since we compared daily data, the buoy data are averaged into daily mean values and the data in each product are spatially interpolated into buoy location.

Table 4. Statistics for each daily product in comparison to buoys.

	001	002	J-OFURO2	HOAPS3	
BIAS [g/kg]	0.46	0.38	0.19	-0.21	
RMSE [g/kg]	1.23	1.01	1.68	1.57	
CORRELATION	0.954	0.970	0.920	0.925	
	Zong	Liu	NRA1	NRA2	JRA25
BIAS [g/kg]	-0.41	0.03	0.72	0.39	0.67
RMSE [g/kg]	1.65	1.88	1.13	1.16	0.95
CORRELATION	0.923	0.906	0.962	0.959	0.973

It can be observed that the biases of the satellite products are smaller than those of the reanalysis products and our products, while the RMS errors and correlation coefficients of the satellite products are larger and lower than those of the reanalysis products and our products, respectively. It is interesting to note that although our products are satellite products, their statistics are similar to those of reanalysis data. Although buoy data are not assimilated into the present products, the RMS errors are significantly small and the correlation coefficients are considerably high for our products. It can be concluded that we can reduce the RMS errors by inclusion of information about the surface air specific humidity in NRA1.

DISCUSSIONS

Figure 1 shows that the buoy distribution is inhomogeneous, for example, many buoys exist in the tropical Pacific regions and to the east of the U.S.A. The statistics given in Table 4 are likely to be affected by this imbalance in the buoy number. We carried out cluster analysis for the daily buoy data for 2004. The results are shown in Figure 3.

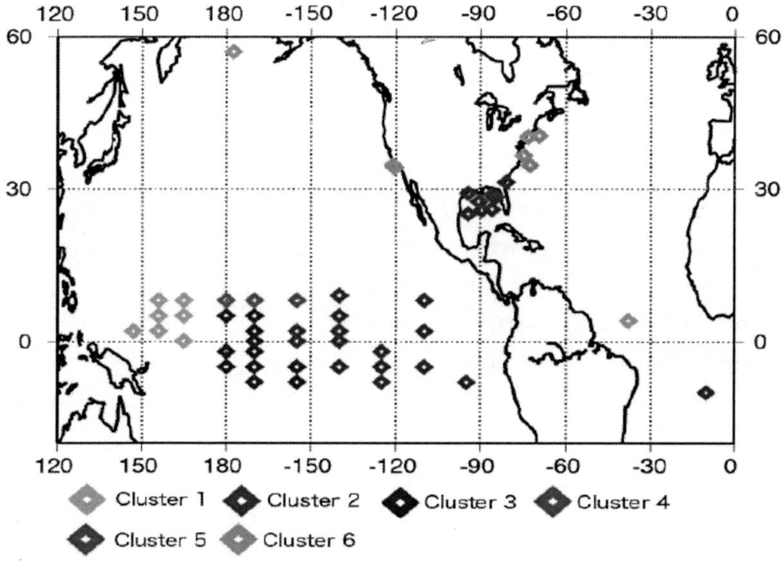

Figure 3. Spatial distribution of buoys in each cluster.

The clusters are separated on the basis of latitudes. However, it is interesting to note that the clusters in the tropics are separated by longitudes. The characteristics of each cluster are not discussed here and will be discussed in a separate study in the near future. After the cluster analysis, we analyzed the values averaged in each cluster in order to remove the imbalance due to the data number. The results are given in Table 5.

Table 5. Statistics for daily products that are averaged in each cluster in comparison to buoy data.

	001	002	J-OFURO2	HOAPS3	
BIAS [g/kg]	0.48	0.38	0.20	-0.20	
RMSE [g/kg]	0.80	0.61	1.17	1.15	
CORRELATION	0.985	0.992	0.967	0.971	
	Zong	**Liu**	**NRA1**	**NRA2**	**JRA25**
BIAS [g/kg]	-0.28	0.04	0.71	0.43	0.83
RMSE [g/kg]	1.31	1.18	0.72	0.77	0.58
CORRELATION	0.967	0.967	0.987	0.985	0.993

The biases in Table 5 are quite similar to those in Table 4, except for that of JRA25. On the other hand, the RMS errors and correlation coefficients in Table 5 show a remarkable decrease and increase in value, in comparison with the products in Table 4, respectively. This means that it is difficult for all data sets to follow the true variation of air specific humidity in the tropical regions. And we understand the accuracy is considerably good except tropical regions.

The algorithm developed by [7], which was developed for AMSR-E data, results in a relatively small bias, shown in Table 5, as compared to the bias for the products obtained by our algorithm. Figure 4 shows a scatter diagram for the KEO data and surface air specific humidity obtained by Zong's algorithm and the present algorithm 002.

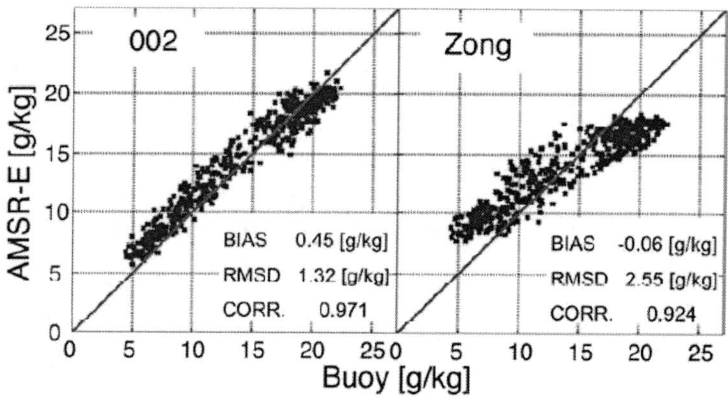

Figure 4. Scatter plots of surface specific humidity derived from KEO buoy and AMSR-E data.

The corresponding statistics are also given in the same figure. The RMS errors were 2.55 g/kg and 1.32 g/kg for Zong's algorithm and the regression formula 002, respectively. From this figure, it can be observed that although the bias for Zong's algorithm is considerably smaller than that for 002, the overestimation of small values of specific humidity offsets the underestimation of large values of specific humidity by Zong's algorithm.

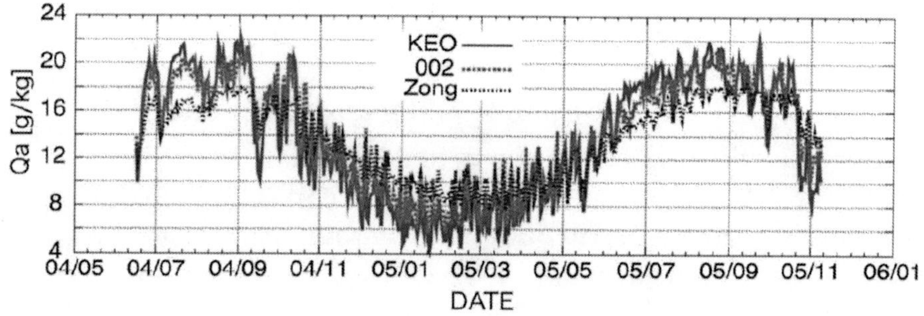

Figure 5. Time variation of surface specific humidity observed by KEO buoy and derived from AME-E data.

Figure 5 shows the time variation of surface air specific humidity. It is clear that Zong's algorithm underestimates the surface air specific humidity in summer and overestimates the surface air specific humidity in winter. The same tendency is found even for 002, but the error is extremely small as compared to Zong' s algorithm. It is concluded that the small bias obtained by using Zong's algorithm is a results of the counterbalance of the underestimation of the specific humidity in summer and overestimation in winter. Thus, it is concluded that a better algorithm could be constructed for the estimation of global surface air specific humidity using AMSR-E data in the present study.

SUMMARY

We developed a new algorithm to determine the surface air specific humidity from the AMSR-E brightness temperature data by assuming the in situ data of ICOADS for the year 2005 to be true values. The use of surface specific humidity data in NRA1 can significantly reduce random errors. We validated the products obtained by the regression formulae by using two kinds of in situ data—ship data and buoy data. The instantaneous surface air specific humidity data obtained by the new algorithm for the period between 2003 and 2004 are compared with in situ data for the same period. We found that the results yielded much better statistical values than those obtained by applying previous algorithms. It should be noted that most previous algorithms were developed not for AMSR-E but for DMSP/SSMI data. The daily specific humidity data was also validated by comparison with the buoy data. It can be observed that the bias is not as small as that for the satellite and reanalysis products.

However, the RMS error is considerably small, i.e.; 0.61 g/kg, as compared to other products. Moreover, the correlation coefficient is considerably high, i.e.; 0.99. Both values are much better than those obtained by the algorithm of [7], which was developed for AMSR-E data. The biases of the products obtained by our algorithms are larger than those of the products obtained by Zong's algorithm for the KEO buoy. However, by observing the time variation of surface air specific humidity observed in Zong' product, we found that the small bias in Zong's product is not homogenous and is caused by an overestimation of the small values of specific humidity in summer and underestimation of the large values of specific humidity in winter Therefore, it can be concluded that the present new algorithm is currently the most superior algorithm to estimate accurate surface air specific humidity using AMSR-E data. Finally the accuracy by the present algorithm is not so different from that by reanalysis data.

However, the accuracy shown by the reanalysis data may be much worse in the regions without in situ data because most of in situ data used in the present validation are assimilated into the reanalysis data.

Therefore, it is important that surface specific humidity having similar accuracy to the reanalysis data near in situ observation points is obtained by the present algorithm constructed by synthesizing satellite and reanalysis data, because the accuracy of satellite data is considerably homogeneous compared with reanalysis data.

ACKNOWLEDGEMENTS

This research was partly funded by the Japan Aerospace Exploration Agency and Tokai University.

REFERENCES AND NOTES

1. Schulz, J.; Schlüssel, P.; Grassl, H. Water vapor in the atmospheric boundary layer over oceans from SSM/I measurements. *Int. J. Remote Sens.* **1993**, 14, 2773–2789.
2. Schlüssel P.; Schanz, L.; Englisch, G. Retrieval of latent heat flux and longwave irradiance at the sea surface from SSM/I and AVHRR measurements. *Adv. Space Res.* **1995**, *16*, 107–116.
3. Chou, S.H.; Atlas, R.M.; Shie, C.L.; Ardizzone, J. Estimates of Surface Humidity and Latent Heat Fluxes over Oceans from SSM/I Data, *Mon. Weather Rev.* **1995,** *123*, 2405–2425.
4. Kubota, M.; Iwasaka, N.; Kizu, S.; Konda, M.; Kutsuwada, K. Japanese Ocean Flux Data Sets with Use of Remote Sensing Observations (J-OFURO). *J. Oceanogr.* **2002**, 58, 213–225.
5. Chou, S.H.; Nelkin, E. Ardizzone, J.; Atlas, R.M.; Shie, C.L. Surface turbulent heat and momentum fluxes over global oceans based on the Goddard Satellite retrievals, version 2 (GSSTF2), *J. Climate* **2003,** *16*, 3256–3273.
6. Grassl H.; Jost, V.; Kumar, R.; Schulz, J.; Bauer, P.; Schluessel, P. *The Hamburg Ocean*-and Fluxes from Satellite Data (HOAPS): A climatological atlas of satellite-derived air-sea-interaction parameters over the oceans; Report No. 312, Max Planck Institute for Meteorology: Hamburg, Germany, 2000.

7. Zong, H.; Liu, Y.; Rong, Z.; Cheng, Y. Retrieval of sea surface specific humidity based on AMSRE satellite data. *Deep-Sea Res. Part I* **2007**, *54*, 1189–1195.

8. Yu, I.; Weller, R.A.; Sun, B. Mean and variability of the WHOI daily latent and sensible heat fluxes at in situ flux measurement sites in the Atlantic Ocean, *J. Climate* **2004**, *17*, 2096-2118.

9. Jiang, C.L.; Cronin, M.F.; Kelly, K.A.; Thompson, L. Evaluation of a hybrid satellite- and MWPbased turbulent heat flux product using Tropical Atmosphere-Ocean (TAO) buoys. *J. Geophys. Res.-Oceans* **2005**, *110*, C09007, doi: 10.1029/2004JC002824.

10. Kalnay, E.C.; Kanamitsu, M.; Kistler, R.; Collins, W.; Deaven, D.; Gandin, L.; Iredell, M.; Saha, S.; White, G.; Woollen, J.; Zhu, Y.; Leetmaa, A.; Reynolds, B.; Chelliah, M.; Ebisuzaki, W.; Higgins, W.; Janowiak, J.; Mo, K.C.; Ropelewski, C.; Wang, J.; Jenne, R.; Joseph, Dennis. The NCEP/NCAR 40-Year Re-analysis Project. *B. Am. Meteorol. Soc.* **1996**, *77*, 437–441.

11. Gulev, S.; Jung, T.; Ruprecht, E. Estimation of the impact of sampling errors in the VOS observations on air-sea fluxes. Part I: Uncertainties in climate means. *J. Climate* **2007**, *20*, 279–301.

12. Fairall, C.W.; Bradley, E.F.; Hare, J.E.; Grachev, A.A.; Edson, J.B. Bulk parameterization of airsea fluxes: Updates and verification for the COARE algorithm, *J. Climate* **2003**, *16*, 571–591.

13. Kubota, M.; Iwabe, N.; Cronin, M. F.; Tomita, H. Surface heat fluxes from the NCEP/NCAR and NCEP/DOE reanalyses at the Kuroshio Extension Observatory buoy site. *J. Geophys. Res. -Oceans,* **2008**, *113*, C02009, doi: 10.1029/2007JC004338.

14. Meindl, E.A.; Hamilton, G.D. Programs of the National Data Buoy Center. *B. Am. Meteorol. Soc.*; **1992,** *73*, 985–993.

15. McPhaden, M.J.; Delcroix, T.; Hanawa, K.; Kuroda, Y.; Meyers, G.; Picaut, J.; Swenson, M. *The El Nino/Southern Oscillation (ENSO) Observing System, Observing the Ocean in the 21st Century*; Australian Bureau of Meteorology: Melbourne, Australia, 2001; pp. 231–246.

16. Liu, W.T. Statistical relation between monthly mean precipitable water and surface-level humidity over global ocean. *Mon. Weather Rev.* **1986**, *14*, 1591–1602.

17. Wilks, D.S. *Statistical methods in the atmospheric sciences, Second edition*; Academic Press: New York, 2006.

18. Bentamy, A.; Katsaros, K.B.; Mestas-Nuñez, A.M.; Drennan, W.M.; Forde, E.B.; Roquet, H. Satellite estimates of wind speed and latent heat flux over the global oceans. *J. Climate* **2003**, 16, 637–656.

19. Singh, R.; Simon, B.; Joshi, P.C. A technique for direct retrieval of surface specific humidity over open oceans from IRS-P4/MSMR satellite. *Bound.-Lay. Meteorol.* **2003**, *106*, 547–559.

20. Jackson, D.L.; Wick, G.A.; Bates, J.J. Near-surface retrieval of air temperature and specific humidity using multisensor microwave satellite observations. *J. Geophys. Res.-Atmos.* **2006**, *111*, (D10306), doi: 10.1029/2005JD006431.

21. Kubota, M.; Tomita, H. The Present State of the J-OFURO Air-Sea Interaction Data Product. *Flux News* **2008**, *4*, 13–15.

22. Kanamitu, M.; Ebisuzaki, W.; Woolen, J.; Yang, S.K.; Hnilo, J.J.; Fiorino, M.; Potter, G.L. NCEPDOE AMIP-II (R-2). *B. Am. Meteorol. Soc.* **2002**, *83*, 1631–1643.

23. Onogi, K.; Tsutsui, J.; Koide, H.; Sakamoto, M.; Kobayashi, S.; Hatsushita, H.; Matsumoto, T.; Yamazaki, N.; Kamahori, H.; Takahashi, K.; Kadokura, S.; Wada, K.; Kato, K.; Oyama, R.; Ose, T.; Mannoji, N.; Taira, R. The JRA-25 reanalysis. *J. Meteorol. Soc. Jpn.* **2007**, *85*, 369–432.

Chapter 3

A FIRST COMPARISON OF SIMULTANEOUS SEA LEVEL MEASUREMENTS FROM ENVISAT, GFO, JASON-1, AND TOPEX/POSEIDON

Caiyun Zhang * and Ge Chen

Key Laboratory of Ocean Remote Sensing, Ministry of Education, Ocean Remote Sensing Institute, Ocean University of China, 5 Yushan Road, Qingdao 266003, China.

E-mails: zhangcy@orsi.ouc.edu.cn, gechen@ouc.edu.cn

ABSTRACT

The multiple altimeter missions have not only advanced our knowledge of ocean circulation, ice sheet topography, and global climate, but also improved the accuracy of altimetric measurements by cross-calibration and validation. In this paper, one year's simultaneous maps of sea level anomaly (MSLA) data obtained from four altimeters, Envisat, Geosat Follow-On (GFO), Jason-1, and TOPEX/Poseidon (T/P), have been compiled for a preliminary comparison. First, the discrepancy in global geographical distribution of each product relative to the merged MSLA field is analyzed and its signal retrieval capability is discussed. Second, the space/time variability of each discrepancy in the Atlantic Ocean, Indian Ocean, Pacific Ocean, Northern Hemisphere, Southern Hemisphere, and global ocean is studied. Third, each discrepancy as a function of latitude, longitude, and merged MSLA is presented. The results show that Jason-1 is the best singlemission for mapping large scale sea level variation, while T/P in its new orbit presents the poorest estimation of SLA due to the short period (from cycle 369 to 403) used to determine the mean profile. A clear understanding of each product discrepancy is necessary for a meaningful combination or merging of multi-altimeter data, optimal product selection, as well as for their assimilation into numerical models.

INTRODUCTION

Sea level change, an important oceanic indicator of climate variation, has been measured by two datasets—tide gauges and satellite altimetry. In recent years, sea level data from altimeters have become much popular in geoscience community as a result of their continuous coverage in space and time, as well as observing capability for both large scale and mesoscale ocean circulation [1-2]. Since 1970s, a series of altimeter missions have been carried out, such as GEOS-3 and Seasat launched in 1975 and 1978, respectively, and Geosat launched in 1986 [3-4]. In 1990s, a new era of space oceanography started with the launch of the European Space Agency (ESA) satellites, ERS-1 and ERS-2, which are for multi-purpose platforms and carrying an altimeter amongst other instruments, and reached its current state-of-the-art by TOPEX/Poseidon (T/P), which was specially designed as an ocean topography mission [5]. For a better monitoring and understanding of long term climate variation, Geosat-Follow-On (GFO), Jason-1, and Envisat, being the follow-on missions of Geosat, T/P, and ERS-2, were successfully launched in 1998, 2001, and 2002, respectively. These multiple altimeter missions have led to vast improvements of accuracy of altimeter measurements by crosscalibration and validation between them, as well as of the mapping capability for mesoscale variability and ocean circulation [6-12].

During 2002-2003, the five altimeters provided sea level data simultaneously with the older missions still in operation while the new follow-on missions were calibrated and validated. Due to technical problems, the ERS-2 tape recorder was switched off in June 2003 and declared to be permanently unavailable in the following month. Before T/P was terminated on 18 January 2006, there were still four altimeters, Envisat, GFO, Jason-1, and T/P, flying concurrently in different orbits and providing operational near-real time sea level products to international users via AVISO. One year's simultaneous data from the four missions are compiled and the purpose of this paper is to reveal the discrepancy of each product compared to the merged MSLA data in order to achieve a better understanding of each product in the mapping capabilities, which has obvious operational relevance, and hopefully the results of this study can be regarded as a reference for users to select suitable products for specific oceanic regions and temporal spans in sea level studies.

ALTIMETER DATA

Weekly near-real time low resolution (1o×1o) of MSLA data obtained from AVISO for the period February 18, 2004 to March 2, 2005 of each mission are compiled in this study. Recent studies have shown that the merged MSLA obtained from multi-missions by use of optimal interpolation really can improve the accuracy of sea level observation and estimation of mesoscale variation (see references 8- 12). Furthermore, recent studies have shown that using such a merged MSLA can reveal some annual amphidromes of sea level variation [13], which are hardly observed by using single missions. Therefore we assume the merged MSLA from four missions as the reference for this study and its RMS is shown in Figure 1. We will compute the RMS difference between the MSLA from individual sensors and the reference MSLA, and refer it as RMS Error (RMSE). A detailed description of MSLA data processing is documented in AVISO user handbook: (M)SLA and (M)ADT near-real time and delayed time products [14]. We are informed by AVISO help (aviso@ cls.fr) that the cycles used for T/P mean profile calculation are started from cycle 369 through cycle 403 (2002.9-2003.8), which is not stated clearly by the user handbook (see reference 14).

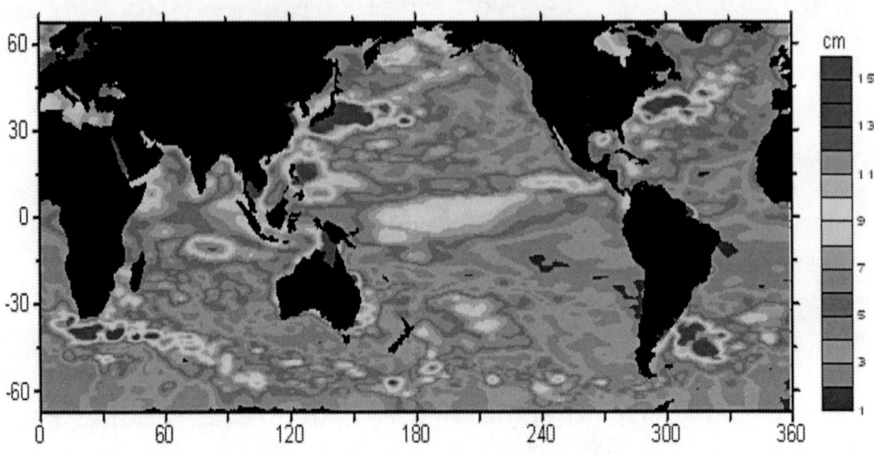

Figure-1. RMS of merged MSLA data based on four altimeters (Envisat, GFO, Jason-1, T/P) between February 18, 2004 and March 2, 2005.

RESULTS

Geographical Patterns

The geographical distributions of RMSE for Envisat, GFO, Jason-1, and T/P are shown in Figure 2. An overall impression is that RMSE of GFO and Jason-1 are much lower than Envisat and T/P. The global mean RMSE over the 55 weeks is 1.429cm, 1.285cm, 1.125cm, and 1.993cm for Envisat, GFO, Jason-1, and T/P, respectively, suggesting that Jason-1 is the best for single-satellite case while T/P having the largest RMSE in its new orbit. As a result of the homogeneity, precision, and consistency of datasets obtained from each mission, the RMSE difference is mostly explained by the mapping error because of the different orbit configuration of each satellite. The mapping discrepancy results for large scale ocean variations obtained from single satellite case are different from mesoscale mapping errors analyzed by Le Traon and Dibarboure [10], and their results show that GFO is the best for mapping mesoscale signals, while Jason-1 has the largest error because T/P and Jason-1 orbits are optimized specially for large scale ocean signal observation. Areas with higher RMSE are mainly found over regions with high ocean variability, such as Antarctic Circumpolar Current (ACC), Gulf Stream, Kuroshio, Brazil/Malvinas Confluence region, and Agulhas Current, where the RMSE varies from 5cm to 25 cm. This confirms that areas with high ocean variability are more difficult to map by use of a single altimeter and that merging has the biggest positive impact on these regions. Part of the higher RMSE over these mesoscale-rich areas might be attributed to the filtering and interpolation involved in the generation of 1-degree products. The large area of high discrepancy of T/P is surprising. We assume that it is mainly caused by the short period (from cycle 369 to cycle 403) mean profile on which SLA is extracted and the large interannual variability of sea level. Since geoids are not known with sufficient accuracy, data along each satellite track need to be referenced to a mean profile (repeattrack analysis) to get SLA measurements. The SLA of Envisat, GFO, and Jason-1 can be extracted based on each mean profile proposed by Le Traon et al. (see reference 15), while for T/P in its new orbit, only one year data are available to calculate the mean profile.

To further examine the mapping capability of each single mission for large scale ocean variation, we present RMSE for each mission as a percentage of RMS of the merged MSLA in Figures 3. By comparing Figures 2 and 3 one can see that regions with low RMSE in Figure 2 (blue color) do not always map the sea level variation very accurately, which can be expected because the 1-degree MSLA data is not designed for mapping the area with very low energy. For Envisat, 27.8% of the whole oceanic gridpoints are clearly different from the merged MSLA, which seems to suggest that the estimations of these gridpoints with percentages more than 30% are poor. As far as GFO and Jason-1 are concerned, estimations are better with the exception of some marginal seas, ACC region from GFO, the south Indian Ocean from Jason-1 single case. The poor mapping results around land may be attributed to the large contamination of tidal aliases over the shallow water. The short time series of T/P data further raises the concern of mapping capability from this newly returned mission.

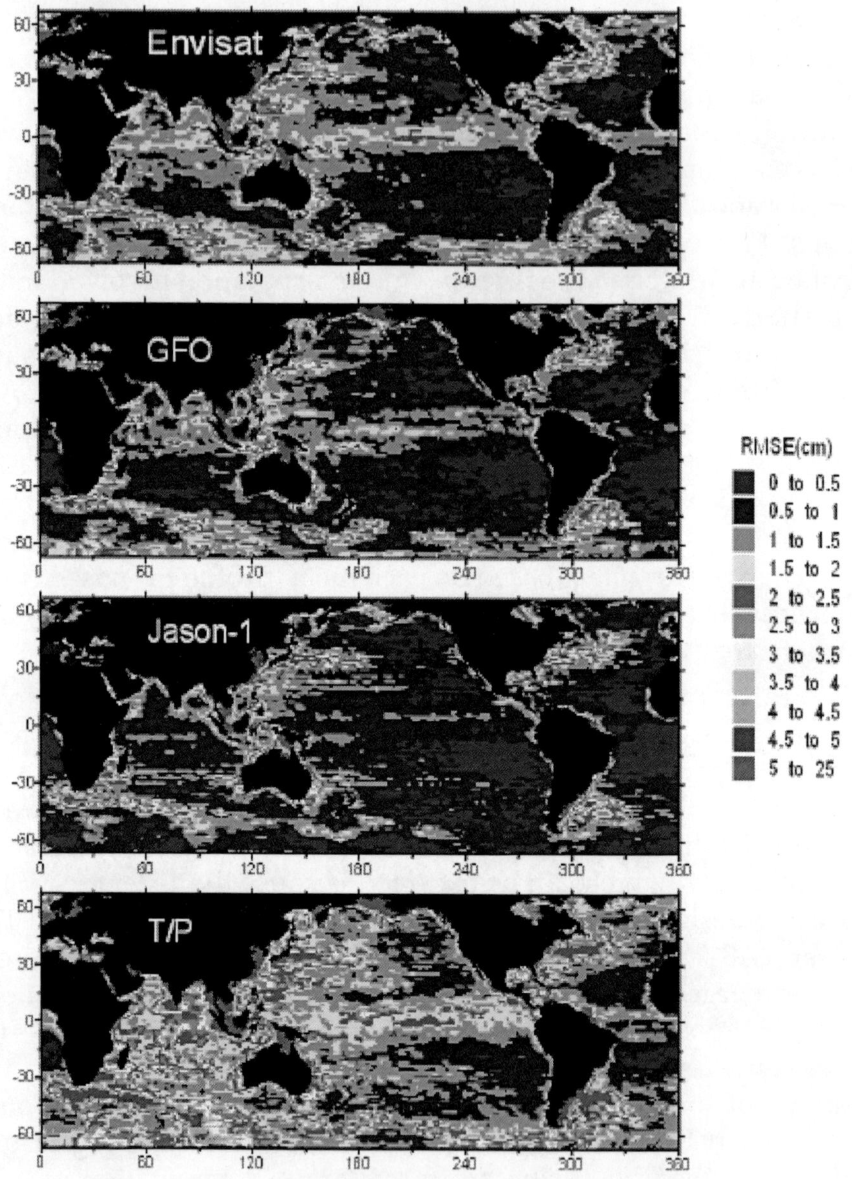

Figure-2. Geographical distributions of RMSE with respect to the merged MSLA for Envisat, GFO, Jason-1, and T/P. The merged MSLA is derived from a combination of simultaneous Envisat, GFO, Jason-1, and T/P estimations from Feb.18, 2004 to Mar. 2, 2005.

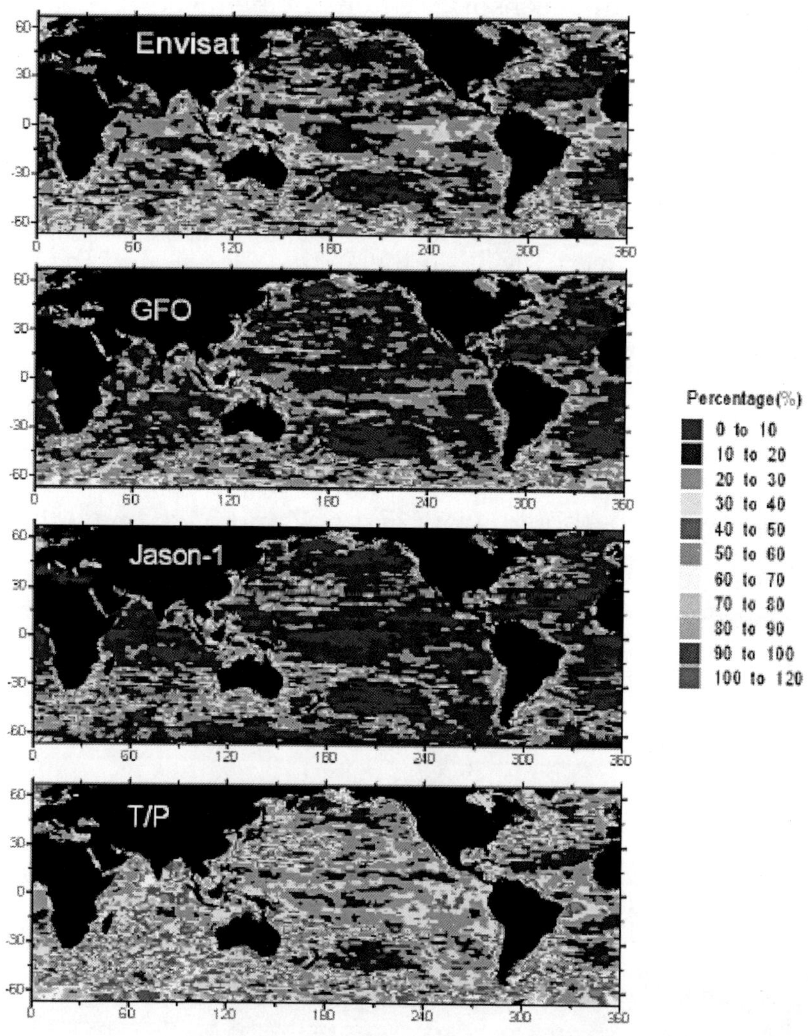

Figure-3. Same as Figure 2 except for being expressed as a percentage of the signal variance.

Spatial Dependence

Figure 2 clearly shows that the same oceanic regions share different RMSE of each sensor and also the same sensor has different RMSE over different regions. For a better understanding this spatial dependence of RMSE of individual sensor, the domain averaged RMSE relative to the merged MSLA for Envisat, GFO, Jason-1, and T/P is given in Figure 4 in terms of the Atlantic Ocean, Indian Ocean, Pacific Ocean, Northern Hemisphere, Southern Hemisphere, and global ocean. The most noticeable feature is that T/P is the poorest case in mapping SLA and Jason-1 is the best single-satellite case from February 2004 to March 2005. Looking at the differences between oceanic basins, each mission has the lowest RMSE in the Pacific Ocean, and the highest RMSE in the Indian Ocean; this difference is particularly pronounced for the T/P case. As far as the two hemispheres are concerned, the Southern Hemisphere (SH) is easier than the Northern Hemisphere (NH) to map because of the larger land distribution in NH. An overall result over the global ocean is that each mission is still suitable to map large scale ocean variation because the largest RMSE, T/P, has not surpassed 3cm, though from Figure 2 and 3 we know that T/P product is the poorest case in its new orbit to map SLA.

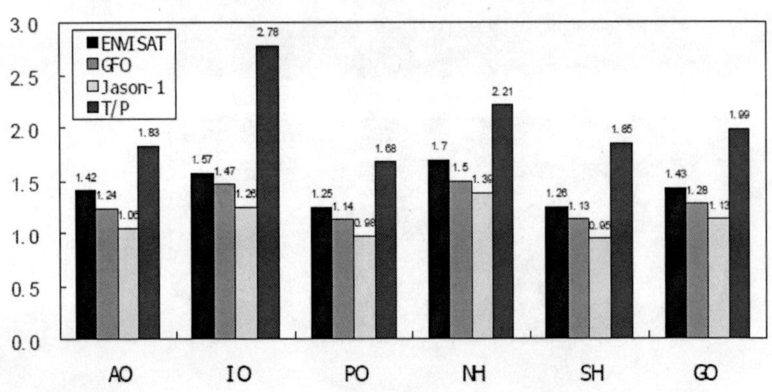

Figure-4. Histogram of RMSE (Unit:cm) with respect to the merged MSLA for Envisat, GFO, Jason-1, and T/P over the Atlantic Ocean (AO), Indian Ocean (IO), Pacific Ocean (PO), Northern Hemisphere (NH), Southern Hemisphere (SH), and global ocean (GO).

Temporal Evolution

One year's simultaneous MSLA is not long enough for us to examine the discrepancy of interannual variation, but its intra-annual variation between February 2004 and March 2005 can be presented.

RMSE evolution for each mission relative to the merged MSLA in the Atlantic Ocean, Indian Ocean, Pacific Ocean, Northern Hemisphere, Southern Hemisphere, and global ocean is shown as in Figures 5. An overall impression is that Envisat, GFO, and Jason-1 are all stable and consistent in time, while T/P is more variable in many regions. In the three oceans and SH, an obvious feature is that GFO and Jason-1 have the lowest RMSE in April, but Envisat seems much higher at the same time. Over the NH and global ocean, it is surprising that Envisat and Jason-1 enjoy a much similar evolution in time with the highest RMSE found in March, while GFO gets its lowest RMSE still in April (Figure 5d, 5e, 5f).

Furthermore, the bias evolution in time at a given location (35oN, 291oE), which is very close to the point (34.8oN, 291.5oE) with high mesoscale variability (in the Gulf Stream) chosen by Le Traon et al. [11], is presented in Figure 6, and we will compare our results with theirs. GFO is the most stable case, while T/P is the most variable one among them. The mean RMSEs at this location for Envisat, GFO, Jason-1, and T/P are 3.53cm, 1.86cm, 3.46cm, and 3.83cm, respectively, while the RMS of sea level signal at the given location is 12.13cm. Thus, such evolution in time will not be a problem forinterpreting the large scale ocean variations.

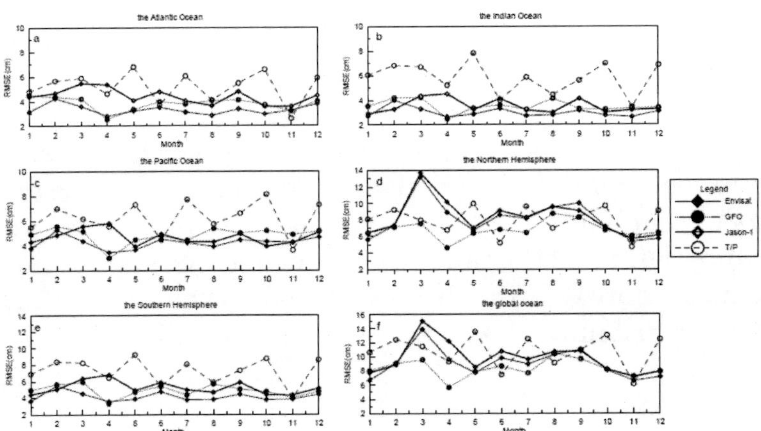

Figure 5. Evolution in time of RMSE relative to the merged MSLA for Envisat, GFO, Jason-1, and T/P at the (a) Atlantic Ocean, (b) Indian Ocean, (c) Pacific Ocean, (d) Northern Hemisphere, (e) Southern Hemisphere, and (f) global ocean.

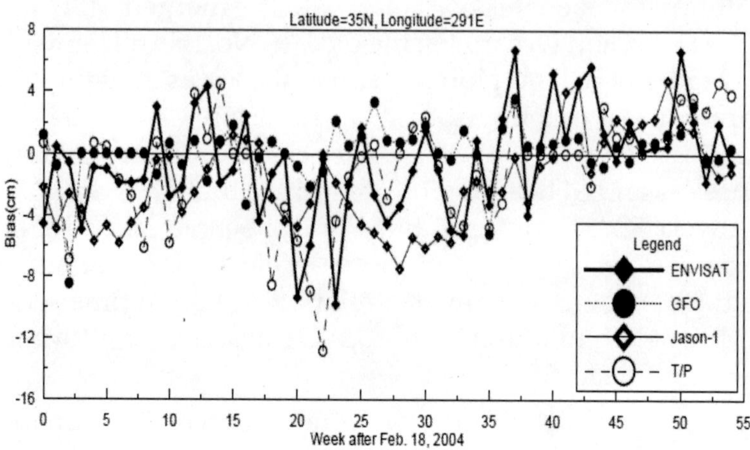

figure 6. Evolution in time of bias relative to the merged MSLA at 35oN, 291oE for Envisat, GFO, Jason-1 and T/P.

It is interesting to look at the phase information of RMSE relative to the merged MSLA and the results are displayed in Figures 7 and 8. It seems that the phase patterns are a little consistent with their corresponding RMSE distributions (Figure 2), with areas of high RMSE showing their maximum(minimum) in March-May (December-February), and those of low RMSE showing theirmaximum (minimum) in December-February (March-May). No obvious transitional phase bands have been found. Except for T/P, the best mapping estimations at most oceanic regions for each mission is in March-May (Figure 8), while from December to next February, the poorest estimations have been presented in mapping (Figure 7). Meanwhile, T/P is out of phase with other three missions in structure, with most oceanic regions being well (poorly) mapped in December-February (March-May). The consistency between phase and RMSE distribution deserves further study.

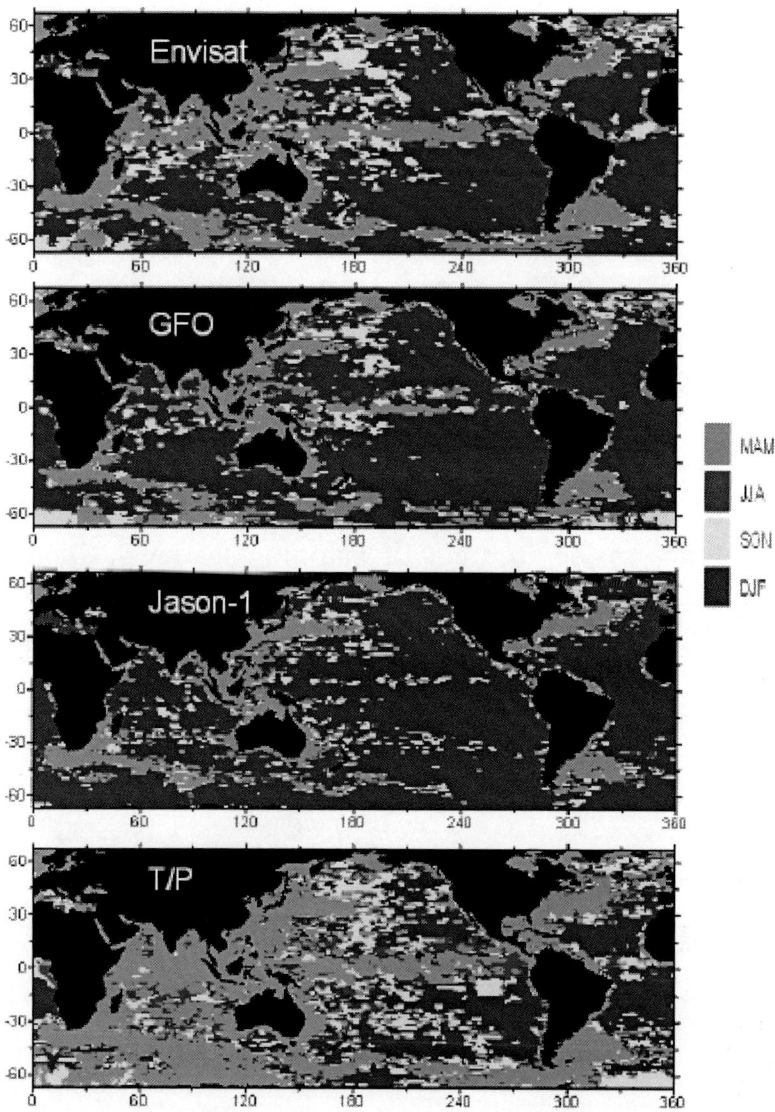

Figure-7 Geographical distributions of timing of maximum RMSE with respect to the merged MSLA for Envisat, GFO, Jason-1, and T/P.

Zonal And Meridional Distribution

RMSE and bias of Envisat, GFO, Jason-1, and T/P are plotted as a function of latitude and longitude in Figure 9. An examination of the meridional variation shows that the four missions have a similar dependence on latitude, with higher RMSE observed over tropical region, mid-latitude, and northernmost latitudes, lower RMSE observed over subtropical regions (Figure 9a). The high RMSE is closely connected with Equatorial currents, prevailing westerlies, and western boundary currents.

The higher values over the northernmost latitudes is surprising, given the higher sampling of each mission of these regions than of low-latitude areas, while over the southernmost latitudes, much lower RMSE is observed. The asymmetry between them suggests that the difference of sea-ice distribution over these two regions has a deep impact on the accuracy of mapping sea level. Clearly note that T/P has the highest RMSE across latitude amongst the four missions, which seems unavoidable because of the short-period used for calculating mean profile of sea level anomaly extraction. The meridional bias seems more stable than the RMSE evolution, except for a large bias which happens around 40oN, where T/P is out of phase of other three missions (Figure 9b).

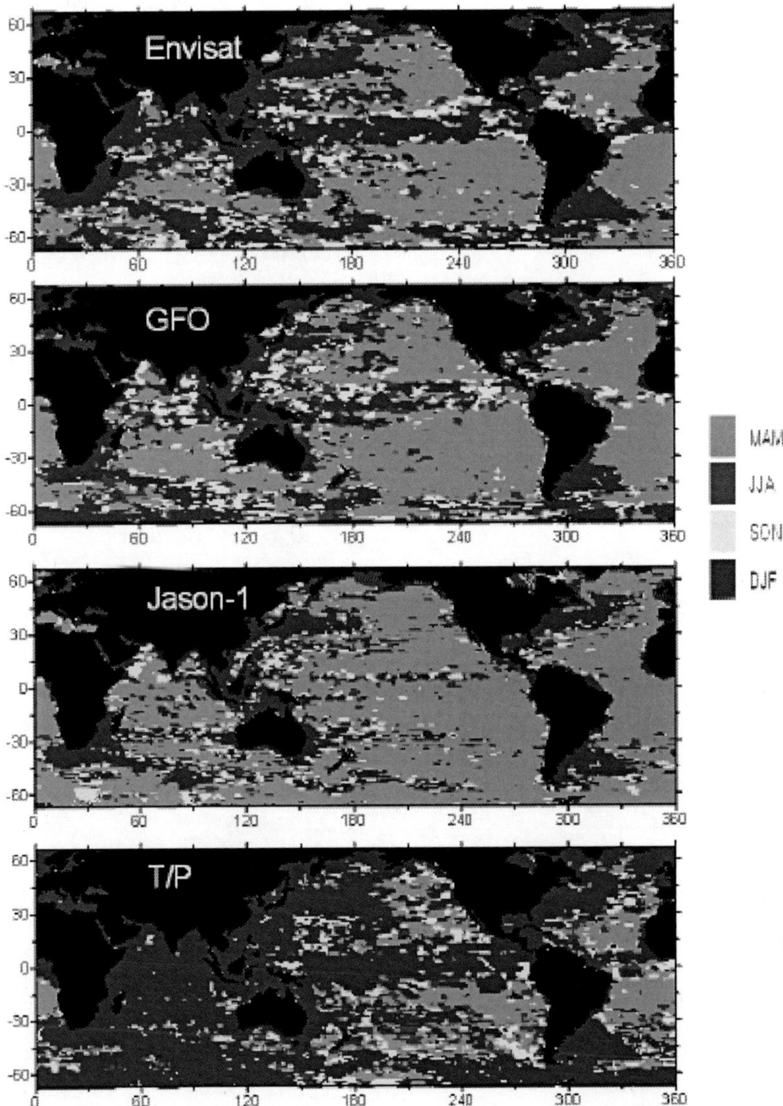

Figure-8. Geographical distributions of timing of minimum RMSE with respect to the merged MSLA for Envisat, GFO, Jason-1, and T/P.

As far as the zonal variation is concerned, the most striking feature is that four peaks appear near 50oE, 120oE, 290oE, and 310oE, respectively (Figure 9c), especially the first spike along 50oE. The last three peaks are related to the Western Boundary Currents, while the first spike, which is far east of the Agulhas Current, may be derived from wrong estimations of four missions, extreme sea level variation in these bands, or few available satellite measurements over these regions to derive the realistic estimate (further scrutiny will be needed to ascertain the precise cause). The bias evolution is also stable, which possibly is derived from the computation of mean bias in time at given locations. However, the out of phase feature between T/P and other three missions around 50oE is also clearly displayed, which suggests that in order to get better mapping measurements at these bands, we can add MSLA of T/P to the average of other three missions to reduce the opposite bias.

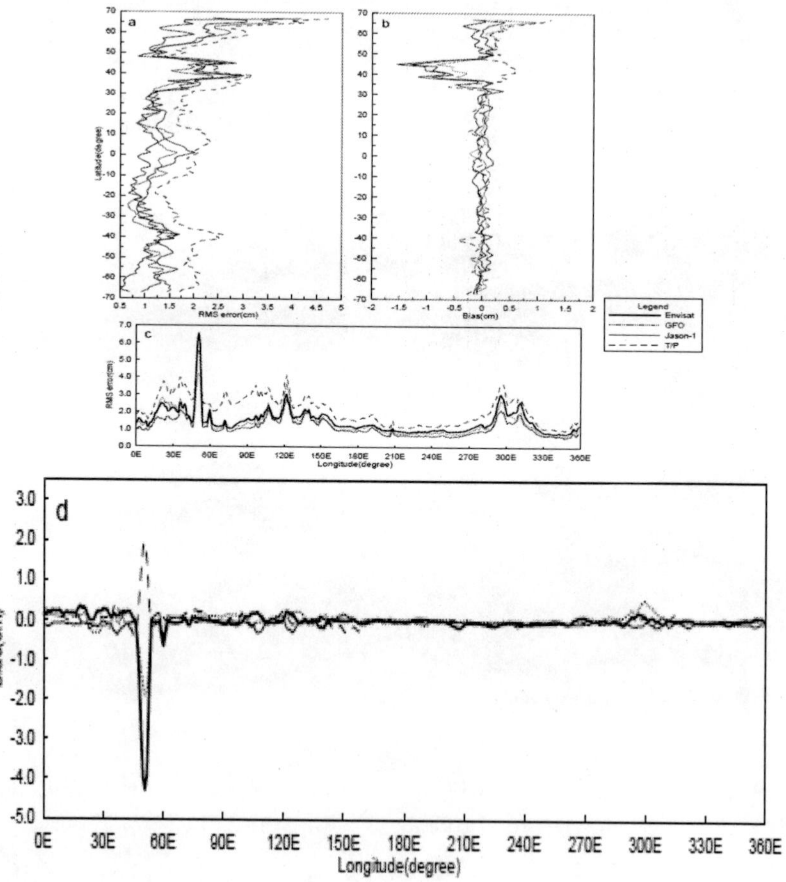

Figure-9. (a), (c) RMSE and (b), (d) bias relative to the merged MSLA as a function of latitudeand longitude.

Dependence on Sea Level Variation

The comparison results have clearly illustrated that the mapping capability of each mission is closely connected with the strength of sea level variation. Thus, RMSE expressed as a percentage and a function of signal variance for Envisat, GFO, Jason-1, and T/P is presented in Figure 10a. The four missions share a very similar trend in signal retrieval. Envisat, GFO, and Jason-1 can map sea level variation well when RMS varies from 2cm to 10cm, and Jason-1 presents the best result, while Envisat displays the poorest estimations. Furthermore, note that most of signal variances are cumulated with RMS around 2-10cm (Figure 10b, the left panel). A decreasing trend in mapping capability has been observed around 10-15cm, and fortunately, only a small amount of gridpoints cumulated within it. To examine the statistical significance for signals with RMSs more than 15cm, we bin RMSs every 2cm in 15-20cm and every 5 cm afterwards, and their associated cumulative plots are presented in Figure 10b (the middle and right panels). It seems that when RMS is more than 17cm, the results presented in the right-hand side of Figure 10a are meaningless due to the small number of gridpoints in each bin.

For a better understanding of the relationship between RMSE and RMS, a histogram plot binned by the signal retrieval is presented in Figure 10c. Note that RMSE expressed as a percentage of RMS of each gridpoint is mainly accumulated within 10%-30%, in which Jason-1 presents the best result, while around 30%-50%, the lowest number of gridpoints of Jason-1 further confirms its mapping capability for large scale of sea level variation. For the best signal retrieval with RMSE only contributing 10% to sea level variation of each mission, Jason-1 still gives the best mapping case due to its largest gridpoints amount among them.

CONCLUSION

Four altimeter missions, Envisat, GFO, Jason-1, and T/P, flying concurrently in orbit and producing operational near-real time sea level data by AVISO since February 2004, provide us with the opportunity to evaluate the mapping capability of large scale sea level variation by each altimeter. Using about one-year's simultaneous

MSLA products between February 18, 2004 and March 2, 2005, a series of statistical comparisons are conducted in this paper. The merged MSLA product from all four altimeters has been selected as the reference field. Our main conclusions are summarized as follows:

(1) Being designed as an optimal measurement for large scale ocean variation, Jason-1 is the best single-satellite mission to produce MSLA data, which can be nicely verified by its much more homogeneous low RMSE in geographical distribution, stable in time evolution, as well as its signal retrieval of sea level variation.

(2) With orbit cycle configured between Jason-1 and Envisat, GFO also enjoys good capability to map the large scale ocean variation and only has a difference of 0.2cm RMSE compared with Jason-1 both in regional and global spatial mean. Furthermore, its time evolution of RMSE and bias is the most stable mission among four missions.

(3) Envisat, being thought to be more suitable to measure the mesoscale ocean variation, indeed has a lower capability to map the large scale sea level variation, which is proved by its large area with higher RMSE, its variable evolution in time, and its larger cumulation for larger RMSE expressed as the percentage of RMS of reference MSLA field (Figure 10c).

(4) Even though RMSE of T/P MSLA product has a similar pattern to the other three missions in time evolution, zonal and meridional variation, and dependence on RMS of reference of MSLA, it still seems that it has the poorest mapping result in its new orbit, which is mainly caused by the inconsistent mean profile of T/P with other three missions.

(5) Because of the homogenous processing of MSLA data, Envisat, GFO, and Jason-1 share much consistency, such as the poor measurement over high variability areas, better (worse) mapping in MAM (DJF) for most oceanic regions, and similar evolution along latitude and longitude, as well as dependence on signal variance.

(6) An interesting feature is that the MSLA estimation is closely connected with seasonal variation.

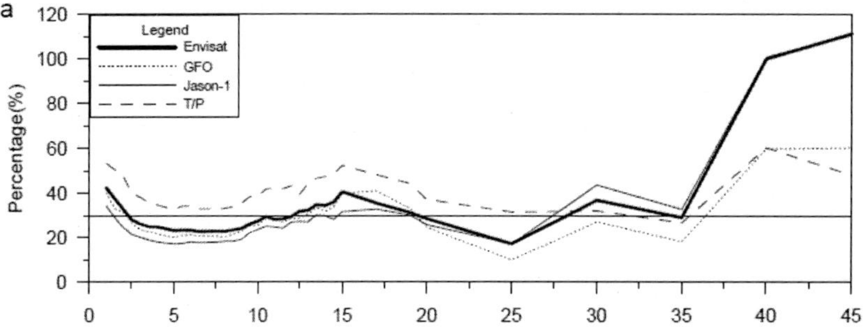

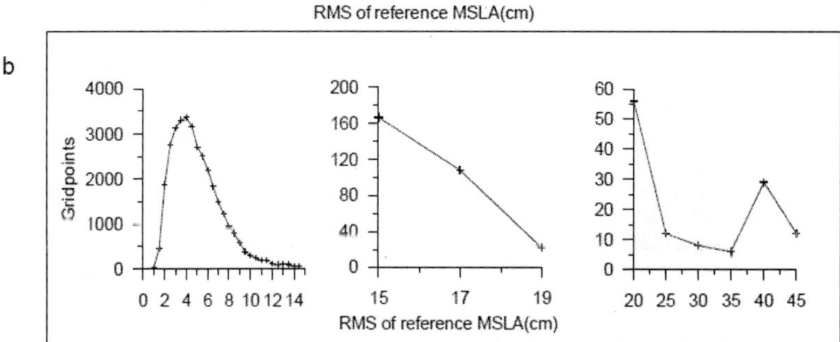

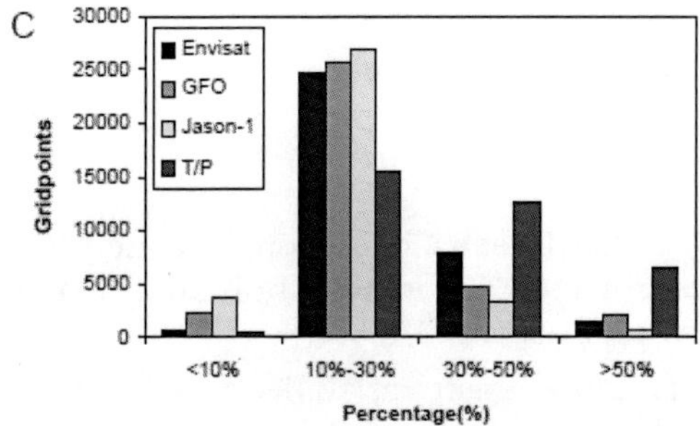

Figure-10. (a) RMSE expressed as percentage of RMS as a function of RMS of reference sea level anomaly for Envisat, GFO, Jason-1, and T/P, respectively. (b) cumulative distribution as a function of RMS, and (c) RMSE histogram binned by its signal retrieval.

The mapping capability for large scale sea level variation of each mission is well measured based on the statistical analysis of RMSE, which provides us with some significant information for the improvement of operational mapping products and optimal selection of each product in applications when one of the missions is, at some time or for some regions, unavailable. The current MSLA product from T/P in its new orbit seems unsuitable for large scale ocean studies, and its application for assimilation into numerical models is not suggested by our results. It is expected that the MSLA data

from T/P in its new orbit will improve once more T/P measurements are used to construct the mean profile, and hopefully its performance would be comparable to the Jason-1 results in the end.

ACKNOWLEDGEMENTS

The authors would like to thank the three anonymous reviewers for their detailed comments and constructive suggestions. We also wish to thank Dr. Zuojun Yu for going through the manuscript and her helpful discussions. This research was jointly supported by the National Basic Research Program of China under Project 2005CB422308, and the Natural Science Foundation of China under Project 40545018.

REFERENCES

1. Fu, L. L.; Chelton, D.; Large-scale Ocean Circulation. Satellite Altimetry and Earth Sciences. A Handbook of Techniques and Application, L-L. Fu and A. Cazenave, Eds., Academic Press, 2001, 133-169.
2. Picaut, J; Busalacchi, A. J.; Tropical Ocean Variability. Satellite Altimetry and Earth Sciences. A Handbook of Techniques and Application, L-L. Fu and A. Cazenave, Eds., Academic Press, 2001, 217-236.
3. Evans, D. L.; Alpers, W.; Cazenave, A.; Elachi, C.; Farr, T.; Glackin, D.; Holt, B.; Jones, L.; Liu, W. T.; McCandless, W.; Menard, Y.; Moore, R.; Njoku, E. Seasat - A 25-year Legacy of Success. Remote Sensing Environment. 2005, 94, 384-404.
4. Douglas, B. C.; Cheney, R. E. Geosat: Beginning a New Era in Satellite Oceanography. J. Geophys. Res. 1990, 95, 2833-2836.

5. Fu, L. L.; Christensen, E. J.; Yamarone Jr, C. A.; Lefevre, M.; Ménard, Y.; Dorrer, M.; Escudier, P. TOPEX/Poseidon Mission Overview. J. Geophys. Res. 1994, 99, 24369-24381.
6. Leuliette, E. W.; Nerem, R. S.; Mitchum, G. T.; Calibration of TOPEX/ Poseidon and Jason Altimeter Data to Construct a Continuous Record of Mean Sea Level Change. Marine Geodesy. 2004, 27, 79-94.
7. Hernandez, F.; Le Traon, P. Y.; Morrow, R. Mapping Mesoscale Variability of the Azores Current Using TOPEX/Poseidon and ERS 1 Altimetry, Together with Hydrographic and Lagrangian Measurements. J. Geophys. Res. 1995, 100, 24995-25006.
8. Le Traon, P. Y.; Nadal, F.; Ducet, N. An Improved Mapping Method of Multisatellite Altimeter Data. J. Atmos. Oceanic Technol. 1998, 15, 522-534.
9. Ducet, N.; Le Traon, P. Y.; Reverdin, G. Global High-resolution Mapping of Ocean Circulation from TOPEX/Poseidon and ERS-1 and –2. J. Geophys. Res. 2000, 105, 19477-19498.
10. Le Traon, P. Y.; Dibarboure, G. Mesoscale Mapping Capabilities of Multi-satellite Altimeter Missions. J. Atmos. Oceanic Technol. 1999, 16, 1208-1223.
11. Le Traon, P. Y.; Dibarboure, G; Ducet, N. Use of High-resolution Model to Analyze the Mapping Sensors 2006, 6 248 Capabilities of Multiple-altimeter Missions. J. Atmos. Oceanic Technol. 2001, 18, 1277-1288.
12. Le Traon, P. Y.; Dibarboure, G. An Illustration of the Contribution of the TOPEX/Poseidon— Jason-1 Tandem Mission to Mesoscale Variability Studies. Marine Geodesy. 2004, 27, 3-13.
13. Chen, G.; Quartly, G. D. Annual Amphidromes: A Common Feature in the Ocean? IEEE Geoscience and Remote Sensing Letters, 2005, 2, 423-427.
14. AVISO. SSALTO/DUACS User Handbook: (M)SLA and (M)ADT Near-real Time and Delayed Time Products. SALP-MU-P-EA-21065-CLS. 2004, CLS.
15. Le Traon, P. Y.; Faugère, Y; Hernandez, F; Dorandeu, J; Mertz, F; Ablain, M. Can We Merge GEOSAT Follow-On with TOPEX/Poseidon and ERS-2 for an Improved Description of the Ocean Circulation? J. Atmos. Oceanic Technol. 2003, 20, 889-895.

Chapter 4

QUALITY VISUALIZATION OF MICROARRAY DATASETS USING CIRCOS

Martin Koch and Michael Wiese*

Pharmaceutical Institute, Rheinische Friedrich Wilhelms University Bonn, An der Immenburg 4, Bonn 53121, Germany; E-Mail: martin.koch@uni-bonn.de

* Author to whom correspondence should be addressed; E-Mail: mwiese@uni-bonn.de; Tel.: +49-228-735-213; Fax: +49-228-737-929.

ABSTRACT

Quality control and normalization is considered the most important step in the analysis of microarray data. At present there are various methods available for quality assessments of microarray datasets. However there seems to be no standard visualization routine, which also depicts individual microarray quality. Here we present a convenient method for visualizing the results of standard quality control tests using Circos plots. In these plots various quality measurements are drawn in a circular fashion, thus allowing for visualization of the quality and all outliers of each distinct array within a microarray dataset. The proposed method is intended for use with the Affymetrix Human Genome platform (*i.e.*, GPL 96, GPL570 and GPL571). Circos quality measurement plots are a convenient way for the initial quality estimate of Affymetrix datasets that are stored in publicly available databases.

Keywords: circos; yaqcaffy; quality monitoring

INTRODUCTION

Microarray Raw Data and Quality Control

Microarray technology has successfully made the transition from a specialized method to a common method that is widely adopted in biological research. However, marginal overlap in gene expression profiling studies [1] and discussions about analysis approaches for identifying differentially expressed genes have prompted a microarray quality initiative. The microarray quality control project (MAQC) aims to estimate the accuracy of the technology [2] and examines in a second phase analysis methods and obtained biological models [3]. In general, the aim of the MAQC-II study was to resolve the concerns about the reproducibility and the generalization capability of microarray analysis results, which may stem from:

(i) a lack of information about the analysis protocol.

(ii) choosing different normalization methods.

(iii) the use of defective analytical methods.

It was found that reproducibility seems to rely greatly on the availability and completeness of the documentation of the analysis process; however it is just as essential to start the analysis with high quality data. Therefore, prior to the analysis of the raw data, the quality of the dataset needs to be assessed. An important goal of quality assessment is detection of outliers. However, currently there is no common quality measure to estimate the soundness of a microarray dataset, although there are a variety of methods available for the analysis of microarray quality in the Bioconductor project [4]. A remedy to this situation is given by post-normalization quality assessment, which detects systematically wrong appliance of normalization. It was earlier suggested that a quality check should be performed before the actual analysis as well, thus preventing the attempt of normalizing erroneous data in the first place [5].

In general normalization methods are specifically designed either for dual channel microarray datasets [6–10] or single channel microarray datasets [11,12]. However, there are also manufacturer dependent normalization methods available for the different flavors of microarray data [12–22].

Aim of the Present Study

Here we present a convenient visualization approach for assessment of individual microarray quality in large data sets using Circos. We have applied state of the art Bioconductor packages, to obtain quality control values. To condense the quality control analysis results, only outliers in the assessed quality measurements are highlighted, *i.e.*, via red dots, denoting absence or deviations of probes in standard technical controls. Additionally RNA degradation assessment is presented in tiles using a gradient from blue to red for depicting potential outlier probes. Finally a principal component plot shows the actual array as a red dot and the other arrays in blue, thus facilitating the detection of potential outlier candidates. Presenting the analysis results in a circular format might be a suitable and convenient way for the initial quality estimate of datasets that are stored in publicly available databases. R code and a demo are publicly available at: https://github.com/buzzmak/circos-arrayQC.

EXPERIMENTAL SECTION

Publicly Available Microarray Studies in GEO

Microarray raw data of publicly available studies was obtained from Gene Expression Omnibus (GEO, [23]). Table 1 lists all studies, which were evaluated for array quality; note that all studies comprise Affymetrix Human Genome U133A arrays and became available recently, *i.e.*, 2011, except for GSE9936 [24] which was submitted to GEO in 2008.

Table 1. The following studies are publicly available as microarray raw data in the Gene Expression Omnibus (GEO) database.

GEO ID	Samples	Reference	Published on
GSE9801	6	[25]	7 November 2011
GSE32700	46	[26]	7 October 2011
GSE9936	105	[24]	7 February 2008

Data Processing, Normalization and Principal Component Analysis

Microarray data handling was performed in R (2.15.0), using the latest version of Bioconductor [4]. Microarray raw data was obtained using the *GEOquery* package [27] and batch processed using the *affy* package [28]. Additionally we used *affy* to calculate potential candidate arrays for RNA degradation. We performed quality control tests using the *yaqcaffy* package, which provides routines that are dedicated to Affymetrix arrays. We calculated all outliers in average background and average noise. Additionally we estimated outliers in both house-keeping probes (*i.e.*, β-Actin and GADPH), as well as outliers in the internal spike-in probe calls and poly-A controls.

Finally all microarrays were processed using quantile normalization of the *RMA* package without background adjustment [18]. Subsequently principal component analysis was performed using the *pcaMethods* package [29]. The scores of the mean centred first principal component were obtained for visualization.

Data Visualization Using Circos

All quality control analysis results were summarized using R and then all dedicated Circos input files were generated by use of the proposed method (R code and a demo are publicly available at: http://github.com/buzzmak/circos-arrayQC).

Inside an R command shell the proposed method can easily be executed, calling the following routine:

```
> writeCircos.files(data, celNames, workdir, fileName, pathToCircos)
```

Where ***data*** presents an *affyBatch* object containing the raw data, ***celNames*** denotes all Affymetrix cel-file names used in the analysis, ***workdir*** is the path to the current directory, ***filename*** names the resulting Circos QC plot and ***pathToCircos*** leads to the home directory of Circos.

Circos [30] version 0.6 and strawberry Perl (http://strawberryperl.com/) version 5.1.16 was used to generate all circular quality plots. Note that the Circos method needs to be called by the Perl interpreter, *i.e.*,

```
> perl bin/circos -conf C:\Path\to\where\circosFiles\are\located\circosQCconfig.txt
```

RESULTS AND DISCUSSION

Quality Assessment and Normalization of Diverse Studies Available in GEO

Since currently the majority of microarray datasets in GEO are based on Affymetrix arrays, we focused our investigation on Human Genome U133A arrays (GPL96 and GPL570) from this platform. Three representative example datasets varying mainly in the number of comprised arrays were selected to show the visualization capability of our approach (Table 1). The principal approach is introduced by using arrays from the study GSE9801. The principal component analysis results, which are made with the *pcaMethods* package [29], are depicted in Figure 1 for this dataset. Here the first principal component is plotted against the second principal component. Two clusters and one potential outlier array can be seen. In Figure 2, a plot of potential RNA degradation is presented, generated with the *affy* package [28]. One array, which represents the topmost line, is probably an outlier. The *yaqcaffy* (http://www.bioconductor.org/packages/release/bioc/html/yaqcaffy.html) package in Bioconductor provides several quality assessment methods for Affymetrix arrays. In Figure 3 there is a quality analysis plot of study

GSE9801 made with *yaqcaffy*. All quality control measurements are demonstrated and the suggested cut-off values are displayed as well. The box-plot showing the GADPH present calls reveals that array GSM247404 is probably an outlier. In addition, the same array is shown also in the Affymetrix specific spike-in and poly-A control plots and therefore most probably presents an outlier. Also there is array GSM247406, which has low values in the dap poly-A control.

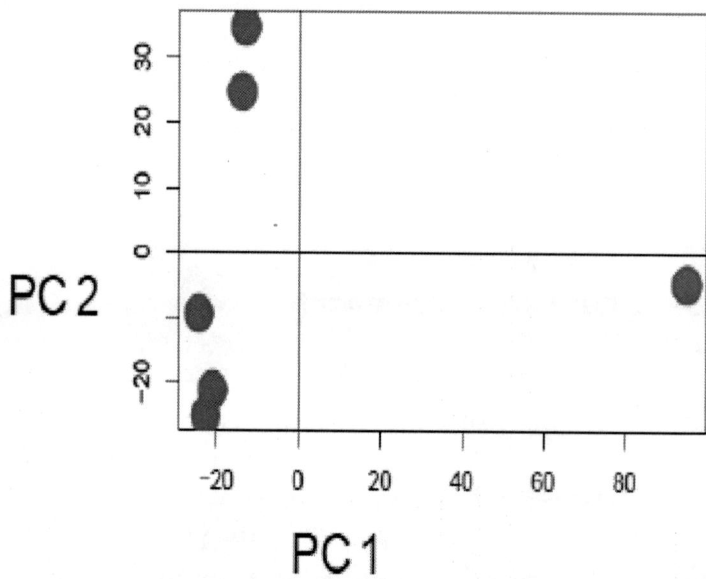

Figure 1. Principal component analysis was performed using the *pcaMethods* package. The figure depicts the first *versus* the second principal component in a scatter plot. There are two clusters and one potential outlier to the right in dataset GSE9801. The first PC explains 67% and the second PC explains 19% of the variance of the data.

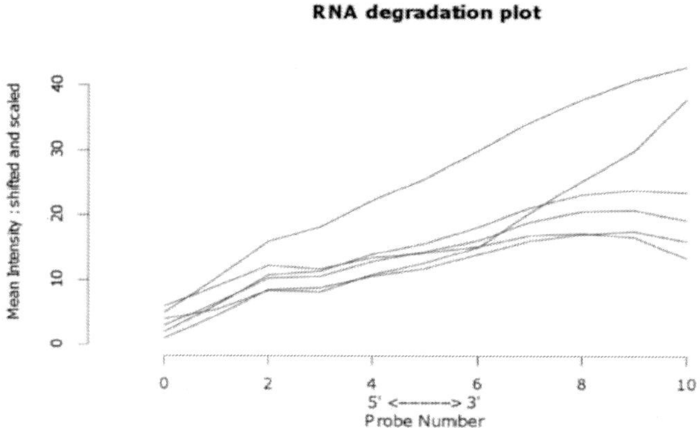

Figure 2. The *affy* package enables the examination of potential RNA degradation probes, *i.e.*, eleven control probes which can reveal a potential fragmentation with high significance. Here we depict potential RNA degradation of arrays in dataset GSE9801. Note that the topmost blue line represents an outlier.

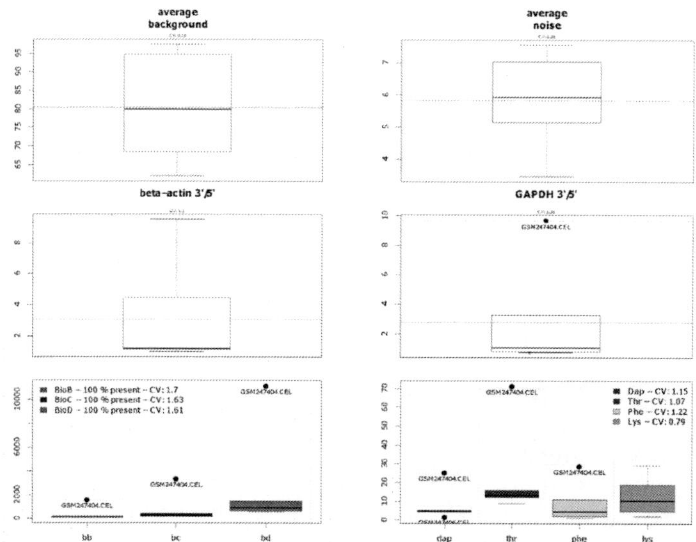

Figure 3. Different quality measurements, which are available in the Bioconductor package yaqcaffy, shown for the example of dataset GSE9801. The first row contains two box-plots, which denote the average background and noise. The second row contains another two box-plots, showing the expression status of so called housekeeping genes. The third row contains box-plots of the Affymetrix specific spike-in and poly-A controls. Note that array GSM247404 is highlighted as potential outlier by black dots. Array GSM247406 is also flagged as outlier in the dap poly-A control, as seen at the bottom right.

Visualization by Use of Circos

In Figure 4 we present an overview of the proposed quality measurement plot, using Circos. The plot combines all previously mentioned quality measurement methods and shows also quality measurements of individual arrays. Here we depict only two arrays, for explanatory purposes. Figure 5 depicts all arrays of a dataset in a circular view. Also we wished to condense quality information almost to the presence of outliers, in this way focusing only on erroneous outlier arrays. The first rim depicts the first principal component and highlights the actual array as a red dot, which is slightly bigger than the blue dots that represent the other arrays. Here arrays having similar principal component scores are clustered together, whereas arrays having different principal component scores are located apart.

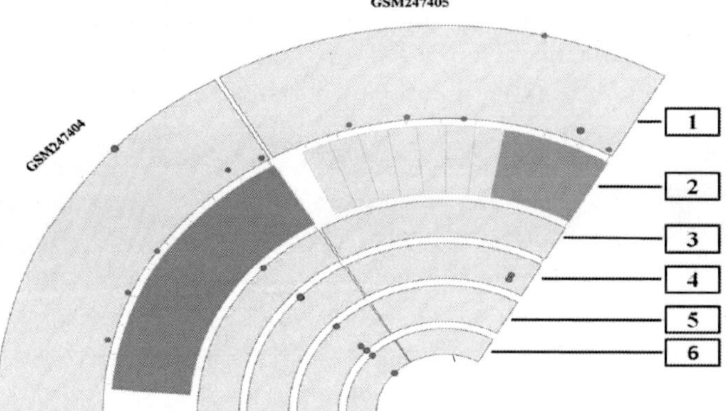

Figure 4. Plot of the different quality measurements, which are shown in Figures 1–3 of the dataset GSE9801 combined with Circos. The outer rim depicts the first principal component, which shows the actual array in red. The second rim displays the result of RNA degradation assessment. Here the red tiles present high values and the blue tiles denote low values of degradation. In addition, white tiles show medium degradation. The next four rims present outliers based on different quality control levels, first average noise and background, second outliers in the level of housekeeping genes and lastly two rims, which are showing outliers in the control spike-in probes. The index in the figure denotes: 1 = first principal component. Note that the actual array is presented as a red dot for better visualisation. 2 = RNA degradation, 3 = background and noise, 4 = β-Actin and GADPH, 5 = internal spike-in probe calls, 6 = poly-A controls. Note that array GSM247404 presents outliers on each level, even when the scores of the first principal component clusters are apart. In contrast to that, the array GSM247405 depicts inconspicuous values

for almost all quality measurements, even GADPH and β-Actin shown on the fourth level in blue appear to be of good quality.

In the next rim potential RNA degradation is depicted. A low overall quality is shown as red tile, while in the case when no degradation was measured the tiles are blue. Medium RNA degradation is depicted in white; these probes are in the range of the tolerance limits. For a better visual inspection of the whole dataset, all arrays are grouped together according to the RNA degradation measure. This way, arrays of similar RNA quality are typically positioned aside in the plot. In the third rim we find red dots if the average background or average noise measurements is off scale. The following three rims are dedicated to several affymetrix specific technical quality control probes and are introduced as follows. The fourth rim comprises the house-keeping genes β-actin and GADPH in blue and flagged in red in case the probes are off scale, *i.e.*, induced more than threefold, these probes will be depicted in red.

The fifth rim presents potential outliers of the internal probe calls, *i.e.*, three probes as depicted in Figure 3 on the bottom left part. The inner ring reveals outliers in poly-A control probes, *i.e.*, maximum of four probes, as depicted in the Figure 3 on the bottom right part. In Figure 4 we present only two arrays from dataset GSE9801, there is one array of low (GSM247404) and another array (GSM247405) of good quality. Array GSM247404 can easily be identified as a potential outlier on all measured quality scales. All technical spike-in probes in rim five are problematic and 50 percent of the other technical controls are outliers as well. The other array (GSM247405) in this dataset however is of good quality, since it shows blue tiles in the RNA degradation plot and no outliers in the technical controls.

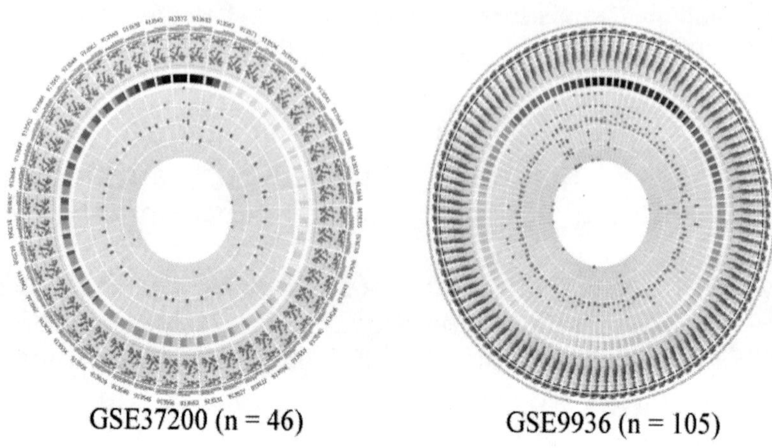

Figure 5. Quality measurement plots of the two datasets listed in Table 1, visualized by Circos, investigating the visualization limit of the proposed method. The plot on the left shows the results of the quality assessment from 46 arrays. The plot on the right summarizes all quality measurements for more than 100 arrays. In this plot one can only trace the quality of the whole dataset; as assessing the quality of single arrays without magnifying is difficult. Therefore we suggest inspecting the same plot in scalable vector format (*i.e.*, svg), as provided in Figure S1 of the supplement.

We wanted to assess the visual capability of our approach and therefore we generated Circos quality plots for two microarray studies as shown in Figure 5. These studies are listed in Table 1 in detail. The left plot in Figure 5 presents all arrays in study GSE37200 comprising 46 arrays and the right plot shows study GSE9936, which contains 105 arrays. In the dataset GSE32700, there are only three arrays of questionable quality. Two of them present outliers, since RNA degradation is detected and there are outliers in average background or noise levels. In addition to this, on the second innermost ring all three spike-in control probes are potential outliers. The majority of the arrays in the dataset are of good quality, there are only occasionally some outliers, mostly in the poly-A controls.

The dataset GSE9936 has several RNA degradation candidate arrays, however only one of these arrays has also additional outliers on all other scales. The other RNA degradation candidates have mostly outliers in the average background intensities. The arrays, where RNA degradation is not problematic show off limit values in average background and noise levels instead. Additionally these arrays have off-scale measurements in both house-keeping control genes and also several absent probes and outliers in the technical spike-in and poly-A controls. In total we can count seven arrays, where quality measurements would imply outlier candidates. This underlines the fact that it might be important to access a plurality of quality measurement methods, to gain outlier arrays with greater confidence. In summary our method suits the majority of the dataset series based on GPL96 and GPL570, since most of these comprise less than 100 arrays. However, we find also that datasets containing more than 100 arrays are not reasonably well pictured using our method. Nonetheless, our method is applicable especially for use in web-resources, since Circos produces scalable vector graphic images in which the area of interest can be easily magnified.

DISCUSSION

The NCBI resource GEO contains at present over 10,000 platforms, of which 1,895 comprise human genomic sequences. Among all samples in GEO (776,566) there are 108,119 samples (13.9%), which are based on the Affymetrix Human Genome platform (*i.e.*, GPL 96, GPL570 and GPL571). Over 13,000 samples are based on the Agilent 4x44k platform (0.94%) and there are 3550 samples based on the Illumina human-6 2.0 expression beadchip platform (0.45%). The proposed Circos method supports quality estimation for the majority of microarray-based experiments, which are based on the Affymetrix platform. However, there are quality-reporting methods for other platforms *i.e.*, beadArray [31] for Illumina data and a proprietary method for Agilent arrays. A standard visualization routine, which also depicts individual microarray quality, could allow for an initial quality estimation of a microarray dataset. In the proposed Circos quality plots we find a promising approach providing quick quality assessment, which could in future be generalized to other microarray platforms as well.

CONCLUSION

Quality control is the most important initial step in the analysis of microarray data. However, until now there has been no standard visualization routine to access individual microarray quality control values of large datasets. Here we present a convenient method for accessing the results of standard quality control test results using Circos. Currently the method works only for Microarray datasets, which are based on the Affymetrix Human Genome platform (*i.e.*, GPL 96, GPL570 and GPL571). In future, this method could be adopted for quality estimation of genomic high throughput data.

ACKNOWLEDGMENTS

We wish to acknowledge the kind help of Martin Krzywinski and Laurent Gatto. Martin Krzywinski did also provide the improved layout of the plots and helped substantially to improve our understanding of the circos method.

REFERENCES

1. Bachtiary, B.; Boutros, P.C.; Pintilie, M.; Shi, W.; Bastianutto, C.; Li, J.-H.; Schwock, J.; Zhang, W.; Penn, L.Z.; Jurisica, I.; *et al.* Gene expression profiling in cervical cancer: An exploration of intratumor heterogeneity. *Clin. Cancer Res.* **2006**, *12*, 5632–5640.
2. Shi, L.; Reid, L.H.; Jones, W.D.; Shippy, R.; Warrington, J.A.; Baker, S.C.; Collins, P.J.; de Longueville, F.; Kawasaki, E.S.; Lee, K.Y.; *et al.* The Microarray Quality Control (MAQC) project shows inter- and intraplatform reproducibility of gene expression measurements. *Nat. Biotechnol.* **2006**, *24*, 1151–1161.
3. Shi, L.; Campbell, G.; Jones, W.D.; Campagne, F.; Wen, Z.; Walker, S.J.; Su, Z.; Chu, T.M.; Goodsaid, F.M.; Pusztai, L.; *et al.* The Microarray Quality Control (MAQC)-II study of common practices for the development and validation of microarray-based predictive models. *Nat. Biotechnol.* **2010**, *28*, 827–838.
4. Gentleman, R.C.; Carey, V.J.; Bates, D.M.; Bolstad, B.; Dettling, M.; Dudoit, S.; Ellis, B.; Gautier, L.; Ge, Y.; Gentry, J.; *et al.* Bioconductor: Open software development for computational biology and bioinformatics. *Genome Biol.* **2004**, doi: 10.1186/gb-2004-5-10-r80.
5. McClure, J.; Wit, E. Post-normalization quality assessment visualization of microarray data. *Comp. Funct. Genomics* **2003**, *4*, 460–467.

6. Yang, Y.H.; Dudoit, S.; Luu, P.; Lin, D.M.; Peng, V.; Ngai, J.; Speed, T.P. Normalization for cdna microarray data: A robust composite method addressing single and multiple slide systematic variation. *Nucleic Acids Res.* **2002**, doi: 10.1093/nar/30.4.e15.

7. Edwards, D. Non-linear normalization and background correction in one-channel cdna microarray studies. *Bioinformatics* **2003**, *19*, 825–833.

8. Xiong, H.; Zhang, D.; Martyniuk, C.J.; Trudeau, V.L.; Xia, X. Using generalized procrustes analysis (gpa) for normalization of cdna microarray data. *BMC Bioinformatics* **2008**, doi: 10.1186/ 1471-2105-9-25.

9. Wu, Y.; Yan, L.; Liu, H.; Sun, H.; Xie, H. A new outlier removal approach for cdna microarray normalization. *BioTechniques* **2009**, *47*, 691–692, 694–700.

10. Wu, Z.; Aryee, M.J. Subset quantile normalization using negative control features. *J. Comput. Biol.* **2010**, *17*, 1385–1395.

11. Geller, S.C.; Gregg, J.P.; Hagerman, P.; Rocke, D.M. Transformation and normalization of oligonucleotide microarray data. *Bioinformatics* **2003**, *19*, 1817–1823.

12. Calza, S.; Valentini, D.; Pawitan, Y. Normalization of oligonucleotide arrays based on the least-variant set of genes. *BMC Bioinforma.* **2008**, doi: 10.1186/1471-2105-9-140.

13. Carvalho, B.; Bengtsson, H.; Speed, T.P.; Irizarry, R.A. Exploration, normalization, and genotype calls of high-density oligonucleotide snp array data. *Biostatistics* **2007**, *8*, 485–499.

14. Rigaill, G.; Hupé, P.; Almeida, A.; Rosa, P.L.; Meyniel, J.-P.; Decraene, C.; Barillot, E. Italics: An algorithm for normalization and DNA copy number calling for affymetrix snp arrays. *Bioinformatics* **2008**, *24*, 768–774.

15. Zeller, G.; Henz, S.R.; Laubinger, S.; Weigel, D.; Rätsch, G. Transcript normalization and segmentation of tiling array data. *Pac. Symp. Biocomput.* **2008**, *538*, 527–538.

16. Autio, R.; Kilpinen, S.; Saarela, M.; Kallioniemi, O.; Hautaniemi, S.; Astola, J. Comparison of affymetrix data normalization methods using 6,926 experiments across five array generations. *BMC Bioinforma.* **2009**, doi: 10.1186/1471-2105-10-S1-S24.

17. Barbacioru, C.C.; Wang, Y.; Canales, R.D.; Sun, Y.A.; Keys, D.N.; Chan, F.; Poulter, K.A.; Samaha, R.R. Effect of various normalization methods on applied biosystems expression array system data. *BMC Bioinforma.* **2006**, doi: 10.1186/1471-2105-7-533.

18. Irizarry, R.; Hobbs, B.; Collin, F.; Beazer-Barclay, Y.D.; Antonellis, K.J.; Scherf, U.; Speed, T.P. Exploration, normalization, and summaries of high density oligonucleotide array probe level data. Biostatistics 2003, 4, 249–264.

19. Wu, W.; Dave, N.; Tseng, G.C.; Richards, T.; Xing, E.P.; Kaminski, N. Comparison of normalization methods for codelink bioarray data. BMC Bioinforma. 2005, doi: 10.1186/1471-2105-6-309.

20. Du, P.; Kibbe, W.A.; Lin, S.M. Lumi: A pipeline for processing illumina microarray. Bioinformatics 2008, 24, 1547–1548.

21. Zahurak, M.; Parmigiani, G.; Yu, W.; Scharpf, R.B.; Berman, D.; Schaeffer, E.; Shabbeer, S.; Cope, L. Pre-processing agilent microarray data. BMC Bioinforma. 2007, doi: 10.1186/1471-2105-8-142.

22. Kerr, K.F. Extended analysis of benchmark datasets for agilent two-color microarrays. BMC Bioinforma. 2007, doi: 10.1186/1471-2105-8-371.

23. Barrett, T.; Suzek, T.O.; Troup, D.B.; Wilhite, S.E.; Ngau, W.C.; Ledoux, P.; Rudnev, D.; Lash, A.E.; Fujibuchi, W.; Edgar, R. NCBI GEO: Mining millions of expression profiles—Database and tools. Nucleic Acids Res. 2005, 33, D562–D566.

24. Chang, E.C.; Charn, T.H.; Park, S.H.; Helferich, W.G.; Komm, B.; Katzenellenbogen, J.A.; Katzenellenbogen, B.S. Estrogen receptors alpha and beta as determinants of gene expression: Influence of ligand, dose, and chromatin binding. Mol. Endocrinol. 2008, 22, 1032–1043.

25. Lutter, D.; Ugocsai, P.; Grandl, M.; Orso, E.; Theis, F.; Lang, E.W.; Schmitz, G. Analyzing m-csf dependent monocyte/macrophage differentiation: Expression modes and meta-modes derived from an independent component analysis. BMC Bioinformatics 2008, doi: 10.1186/1471-2105-9-100.

26. Aoyagi, K.; Minashi, K.; Igaki, H.; Tachimori, Y.; Nishimura, T.; Hokamura, N.; Ashida, A.; Daiko, H.; Ochiai, A.; Muto, M.; et al. Artificially induced epithelial-mesenchymal transition in surgical subjects: Its implications in clinical and basic cancer research. PLoS One 2011, 6, e18196.

27. Sean, D.; Meltzer, P.S. GEOquery: A bridge between the Gene Expression Omnibus (GEO) and bioconductor. Bioinformatics 2007, 23, 1846–1847.

28. Gautier, L.; Cope, L.; Bolstad, B.M.; Irizarry, R.A. Affy—Analysis of affymetrix genechip data at the probe level. Bioinformatics 2004, 20, 307–315.

29. Stacklies, W.; Redestig, H.; Scholz, M.; Walther, D.; Selbig, J. Pcamethods—A bioconductor package providing pca methods for incomplete data. Bioinformatics 2007, 23, 1164–1167.

30. Krzywinski, M.; Schein, J.; Birol, I.; Connors, J.; Gascoyne, R.; Horsman, D.; Jones, S.J.; Marra, M.A. Circos: An information aesthetic for comparative genomics. Genome Res. 2009, 19, 1639–1645.

31. Ritchie, M.E.; Dunning, M.J.; Smith, M.L.; Shi, W.; Lynch, A.G. BeadArray expression analysis using bioconductor. PLoS Comput. Biol. 2011, 7, e1002276.

APPENDIX

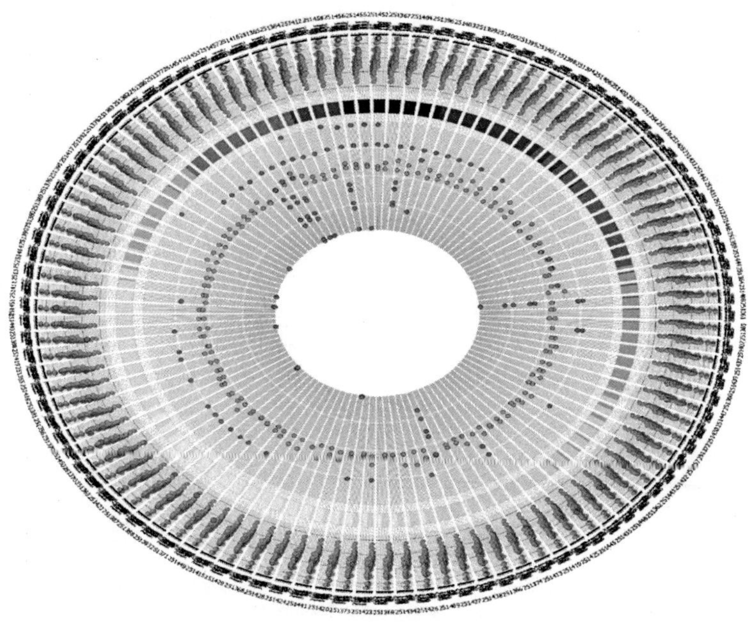

Figure S1. Larger version of the quality measurement plots of data-set GSE9936 as visualized by Circos.

Chapter 5

DISTRIBUTED PRINCIPAL COMPONENT ANALYSIS FOR WIRELESS SENSOR NETWORKS

Yann-A¨el Le Borgne [1*], Sylvain Raybaud[2], and Gianluca Bontempi[1]

[1] Machine Learning Group, D´epartement d'Informatique, Facult´e des Sciences, Universit´e Libre de Bruxelles, Boulevard du Triomphe, 1050 Brussels, Belgium

[2] ´ Ecole Normale Sup´erieure de Cachan, 61, Avenue du Pr´esident Wilson, 94235 Cachan Cedex, France E-mails: yleborgn@ulb.ac.be; sraybaud@dptmaths.ens-cachan.fr; gbonte@ulb.ac.be

[*] Author to whom correspondence should be addressed.

ABSTRACT

The Principal Component Analysis (PCA) is a data dimensionality reduction technique well-suited for processing data from sensor networks. It can be applied to tasks like compression, event detection, and event recognition. This technique is based on a linear transform where the sensor measurements are projected on a set of principal components. When sensor measurements are correlated, a small set of principal components can explain most of the measurements variability. This allows to significantly decrease the amount of radio communication and of energy consumption. In this paper, we show that the power iteration method can be distributed in a sensor network in order to compute an approximation of the principal components. The proposed implementation relies on an aggregation service, which has recently been shown to provide a suitable framework for distributing the computation of a linear transform within a sensor network. We also extend this previous work by providing a detailed analysis of the computational, memory, and communication costs involved. A compression experiment involving real data validates the algorithm and illustrates the tradeoffs between accuracy and communication costs.

INTRODUCTION

Efficient in-network data processing is a key factor for enabling wireless sensor networks (WSN) to extract useful information and an increasing amount of research has been devoted to the development of data processing techniques [1–4]. Wireless sensors have limited resource constraints in terms of energy, network data throughput and computational power. In particular, the radio communication is an energy consuming task and is identified in many deployments as the primary factor of sensor node's battery exhaustion [5]. Emitting or receiving a packet is indeed orders of magnitude more energy consuming than elementary computational operations. The reduction of the amount of data transmissions has therefore been recognized as a central issue in the design of wireless sensor networks data gathering schemes [6]. Data compression is often acceptable in real settings since raw data collected by sensors typically contain a high degree of spatio-temporal redundancies [5, 7–9]. In fact, most applications only require approximated or high-level information, such as the average temperature in a room, the humidity levels in a field with a 10% accuracy, or the detection and position of a fire in a forest.

An attractive framework for processing data within a sensor network is provided by the data aggregation services such as those developed at UC Berkeley (TinyDB and TAG projects) [10, 11], Cornell University (Cougar) [12], or EPFL (Dozer)[13]. These services aim at aggregating data within a network in a time- and energy-efficient manner. They are suitable when the network is connected to a base station from which queries on sensor measurements are issued. In TAG or TinyDB, for example, queries are entered by means of an SQL-like syntax which tasks the network to send raw data or aggregates at regular time intervals. These services make possible to compute "within the network" common operators like average, min, max, or count, thereby greatly decreasing the amount of data to be transmitted. Services typically rely on synchronized routing trees along which data is processed and aggregated along the way from the leaves to the root [10, 11].

Recently, we have shown that a data aggregation service can be used to represent sensor measurements in a different space [14]. We suggested that the space defined by the principal component basis, which makes data samples uncorrelated, is of particular interest for sensor networks. This basis is returned by the Principal Component Analysis (PCA) [15], a well-known technique in multivariate data analysis. The design of an aggregation scheme which distributes the computation of the principal component scores (i.e., the transformed data in the PCA space) has three major benefits. First, the PCA provides varying levels of compression accuracies, ranging from constant approximations to full recovery of original data. Second, simple adaptive protocols can leverage this flexibility by trading network resources for data accuracy. Third, principal component scores contain sufficient information for a variety of WSN applications like approximate monitoring [16], feature prediction [17, 18] and event detection [19, 20].

The approach we proposed in [14] exclusively addresses the distribution of the computation of the principal component scores and requires the component basis to be computed beforehand in a centralized manner. This limits the applicability of the PCA to small networks, as the centralized computation of the principal component basis does not scale with the network size.

The main contribution of this article is to provide a distributed implementation of the principal com ponent basis computation. The algorithm is based on the Power Iteration Method [21, 22], an iterative technique for computing the principal component basis, which can be implemented in a distributed manner by means of an aggregation service. In particular we show that this algorithm can properly compute the principal component basis under the mild hypothesis that the sensor measurements collected by distant sensors are not significantly correlated. We also extend the previous work of [14] by an in-depth discussion of the network tradeoffs and scalability issues.

The article is structured as follows. Section 2 reviews previous work in the domain by describing the principles of data aggregation services in WSN and detailing how principal component scores can be computed by an aggregation service. This section also provides a brief overview of PCA together with potential WSN applications and related tradeoffs. Section 3 presents the main issues in the distribution of the PC basis and proposes an implementation based on a data aggregation service. Experimental results illustrating the tradeoffs between accuracy and communication costs are discussed in Section 4. Additional related work and future research tracks are addressed in Section 5.

Principal Component Aggregation in Wireless Sensor Networks

Let us consider a static network of size p whose task is to collect at regular time instants sensor measurements or aggregates thereof, and to route them to a destination node. This scheme is standard for plenty of real-time WSN applications like surveillance, actuator control or event detection. The destination node is commonly referred to as the sink or the base station and is often assumed to benefit from higher resources (e.g., a desktop computer). Let $t \in \mathbb{N}$ refer to a discretized time domain and xi[t] be the measurement collected by sensor *i*, $1 \leq i \leq p$, at time t. At each time instant t, the p resulting measurements can be seen as components of a vector $x[t] \in Rp$. The sampling period is also referred to as *epoch*.

Since the communication range of a single node is limited, sensors which are not in communication range of the sink have their measurements relayed by intermediate sensors. A standard approach consists in setting up a multi-hop network by means of a routing tree whose root is connected to the sink (Figure 1). Different metrics such as hop count, latency, energy consumption and network load may be taken into account during the design of the routing tree [23].

Data Aggregation

Aggregation Service

An aggregation service allows to aggregate data within a network in a time- and energy-efficient manner. A well-known example of aggregation service is the TAG system, developed at the University of Berkeley, California [10, 11]. TAG stands for Tiny AGgregation and is an aggregation service for sensor networks which has been implemented in TinyOS, an operating system with a low memory footprint specifically designed for wireless sensors [24]. In TAG, an epoch is divided into time slots so that sensors' activities are synchronized according to their depth in the routing tree. Any algorithm can be relied on to create the routing tree, as long as it allows data to flow in.

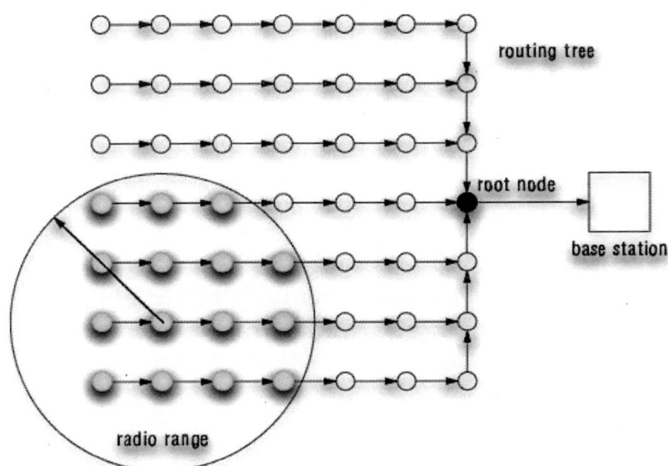

Figure 1. A multi-hop network where all the sensor nodes are connected to the base station by means of a routing tree. Note that as the radio range usually allows a node to communicate with several others, different routing trees can be obtained. The black node represents the root node.

The TAG service focuses on low-rate data collection tasks which permits loose synchronization of the sensor nodes. The overhead implied by the synchronization is therefore assumed to be low. The goal of synchronization is to minimize the amount of time spent by sensors in powering their different components and to maximize the time spent in the idle mode, in which all electronic components are off except the clock. Since the energy consumption is several orders of magnitude lower in the idle mode than when the CPU or the radio is active, synchronization significantly extends the wireless sensors' lifetime.

An illustration of the sensors' activities during an epoch is given in Figure 2 for a network of four nodes with a routing tree of depth three. Note that the synchronization is maintained at the transport layer of the network stack, and does not require precise synchronization constraints as in TDMA. Rather, nodes are synchronized by timing information included in data packets, and a CSMA-like MAC protocol with a random backoff scheme is used at the link layer to resolve multiple access collisions.

Aggregation primitives

Once the routing tree is set up and the nodes synchronized, data can be aggregated from the leaves to the root. Each node adds its contribution to a partial state record X which is propagated along the routing tree. Partial state records are merged when two (or more) of them arrive at the same node. When the partial state record is eventually delivered by the root node to the base station, the desired result is returned by means of an evaluator function. An aggregation service requires then the definition of three primitives [10, 11]:

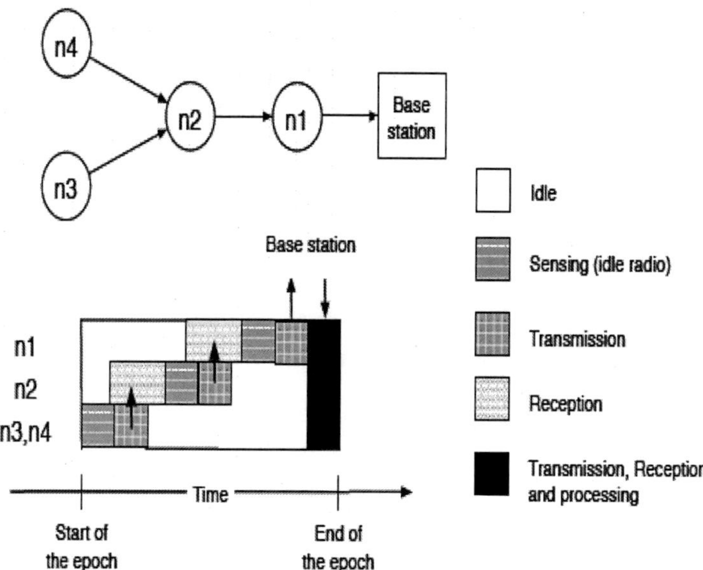

Figure 2. Activities carried out by sensors in the routing tree (adapted from [13]). Transmissions are synchronized for optimizing energy savings. The last stage involves all sensors and allows unsynchronized operations (for sensor discovery e.g.).

- an initializer init which transforms a sensor measurement into a partial state record,
- an aggregation operator f which merges partial state records, and
- an evaluator e which returns, on the basis of the root partial state record, the result required by the application.

Note that when partial state records are scalars or vectors, the three operators defined above may be seen as functions. Partial state records may however be any data structure which, following the notations of [10], are represented using the symbols (.).

We illustrate the aggregation principle by the following example. Suppose that we are interested in computing the Euclidean norm of the vector containing the WSN measurements at a given epoch. Rather than by directly sending all the p measurements to the base station for computation, an aggregation service (Figure 3) can obtain the same result in an online manner once the following primitives are implemented :

$$
\begin{aligned}
init(x) &= \langle x^2 \rangle \\
f(\langle X \rangle, \langle Y \rangle) &= \langle X + Y \rangle \\
e(\langle X \rangle) &= \sqrt{X}
\end{aligned}
$$

In this example the partial state records X and Y are scalars of the form

$$X = \sum_{i \in I} x_i^2$$

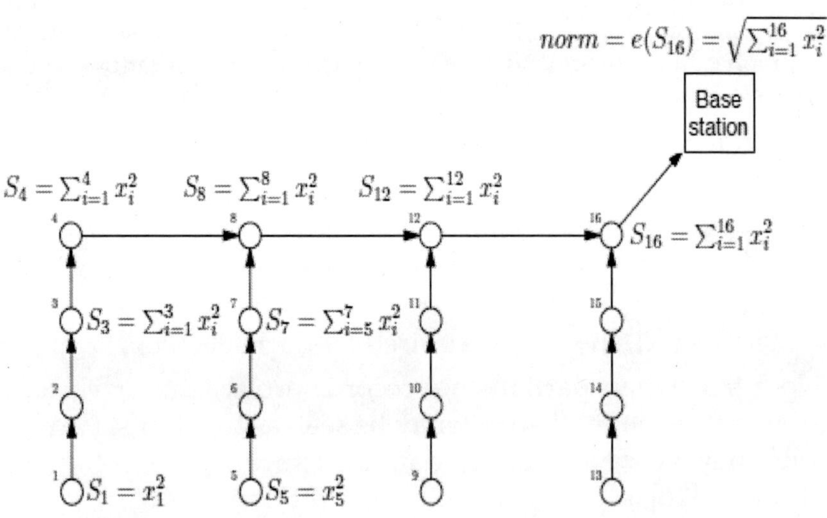

Figure 3. Aggregation service at work for computing the norm of the vector measured. where I $\subset$ f1; : : : ; pg.

It is worth noting that in-network aggregation does not only reduce the amount of packets transmitted but also increases the network lifetime. The energy consumption of the processing unit of a sensor node is at least one order of magnitude lower than the one of the wireless radio [6, 25, 26]. According to the analysis [27] of the main wireless sensor platforms available today on the market, sending one bit of data is equivalent to 2000 CPU cycles in terms of energy. For a packet length of 30 bytes (average TinyOS packet length), this corresponds to 480000 CPU cycles. It follows that, with respect to transmission costs, the energy consumption of simple aggregation operators is negligible. There is therefore a large variety of in-network processing algorithms that can lead to energy savings by reducing the number of packet transmissions.

Communication Costs

This section details the communication costs caused by three types of network operation. The first is the default data collection operation in which all measurement are routed to the sink without any aggregation. This is referred to as the D operation. The second is the aggregation operation, referred to as A operation, which consists in tasking the network to retrieve an aggregate by means of the aggregation service. Finally, we denote by F the feedback operation which consists in flooding the aggregate obtained at the sink back to the whole set of sensors. Let q be the size of a partial state record, Ci be the number of direct children of node i in the routing tree, RTi be the size of the subtree whose root is the i-th node and i C = arg maxi Ci the node whose number of children is the highest.

The following analysis compares the orders of magnitude of the communication costs caused by the D, A and F operations, respectively. For this reason we consider the number of packets processed by each node in an ideal case where overhearing, collisions or retransmissions are ignored.

D operation Without aggregation, all the measurements are routed to the base station by means of the routing tree. As mentioned before, the network load at the sensor nodes, i.e., the sum of received and transmitted packets, is ill-balanced. The load is the lowest at by leaf nodes, which only send one packet per epoch, while the load is the highest at the root node which processes 2p�1 packets (p�1 receptions and p transmissions) per epoch. The load at a generic i-th sensor node depends on the routing tree, and amounts to 2RTi ? 1 packets per epoch.

A operation During the aggregation, the i-th node sends q packets and receives a number of packets which depends on its number Ci of children. The total number of packets processed is therefore q(Ci+1) per epoch. The load is the lowest at leaf nodes, which only have q packets to send, while the load is the highest at the node whose number of children is the highest.

F operation The feedback operation consists in propagating the aggregated value back from the root down to the all leaves of the tree. This operation can be used, for instance, to get all sensor nodes acquainted with the overall norm of their measurements or with the approximation evaluated at the sink. The feedback of a packet containing the result of the evaluation generates a network load of two packets for all non-leaf nodes (one reception and one transmission for forwarding the packet to the children) and of one packet for the leaves (one reception only).

Principal Component Analysis

This section describes the Principal Component Analysis (PCA), a well-known dimensionality reduction technique in statistical data analysis, data compression, and image processing [28, 29]. The technique is based on a basis change which reduces the number of coordinates used to represent the data while maximizing the variance retained. Its use is particularly adapted to correlated data where the dimensionality of the data can be strongly reduced while retaining most of the measurement variations. Given a set of N centered multivariate measurements x[t] 2 Rp used as training samples, 1 t N, the PCA basis is obtained by minimizing the following optimization function

$$\begin{aligned} J_q(\mathbf{x}[t], \mathbf{w_k}) &= \frac{1}{N}\sum_{t=1}^{N} \|\mathbf{x}[t] - \sum_{k=1}^{q} \mathbf{w_k}\mathbf{w_k^T}\mathbf{x}[t]\|^2 \\ &= \frac{1}{N}\sum_{t=1}^{N} \|\mathbf{x}[t] - \hat{\mathbf{x}}[\mathbf{t}]\|^2 \end{aligned} \quad (1)$$

where the set of q p vectors fwkg1kq of Rp form the PCA basis. The linear combinations ^x[t] = Pq k=1 wkwk Tx[t] represent the projections of x[t] on the PCA basis, and Eq. (1) therefore quantifies the mean squared error the original measurements x[t] and their projections. Note that the optimization function equivalently maximizes the variance retained by the family of vectors fwkg1kq. Setting to

where $\mathbf{C} = \frac{1}{N}\sum_{t=1}^{N}\mathbf{x}[t]\mathbf{x}[t]^T$ is the sample covariance matrix of the sensor measurements x[t]. The minimizer of Equation (1) is therefore given by the set of the first q eigenvectors fwkg of the sample covariance matrix C [28]. These eigenvectors are called the principal components and form the PCA basis. The corresponding eigenvalues k quantify the amount of variance conserved by the eigenvectors. Indeed, left-multiplying Equation (2) by w_k^T gives

$$\mathrm{w_k}^T \mathrm{C}\mathrm{w_k} = \mathrm{w_k}^T \lambda_k \mathrm{w_k} \tag{3}$$

and as

$$\mathrm{w_k}^T \mathrm{C}\mathrm{w_k} = \mathrm{w_k}^T \left(\frac{1}{N}\sum_{t=1}^{N}\mathrm{x}[t]\mathrm{x}[t]^T\right)\mathrm{w_k} = \sum_{t=1}^{N}\hat{\mathrm{x}}[t]^T\hat{\mathrm{x}}[t] \text{ and } \mathrm{w_k}^T\mathrm{w_k} = 1$$

each eigenvalue λk quantifies the variance of the projections of the measurements x[t] on the corresponding k-th eigenvector.

The sum of the eigenvalues therefore equals the total variance of the original set of observations X, i.e.: each eigenvalue λ_k quantifies the variance of the projections of the measurements x[t] on the corresponding k-th eigenvector.

The sum of the eigenvalues therefore equals the total variance of the original set of observations X, i.e.:

$$\sum_{k=1}^{p} \lambda_k = \frac{1}{N} \sum_{t=1}^{N} ||\mathbf{x}[t]||^2$$

It is convenient to order the vectors W_k by decreasing order of the eigenvalues, so that the proportion P of retained variance by the first q principal components can be expressed by:

$$P(q) = \frac{\sum_{k=1}^{q} \lambda_k}{\sum_{k=1}^{p} \lambda_k} \qquad (4)$$

Storing columnwise the set of vectors $\{\mathbf{w_k}\}_{1 \le k \le q}$ in a $W_{p \times q}$ matrix, approximations $\hat{\mathbf{x}}[\mathbf{t}]$ to $\mathbf{x}[t]$ in $\mathbb{R}^p$ are obtained by:

$$\hat{\mathrm{x}}[\mathrm{t}] = WW^T\mathrm{x}[t] = W\mathrm{z}[\mathrm{t}] \qquad (5)$$

Where

$$\mathrm{z}[\mathrm{t}] = W^T\mathrm{x}[t] = \begin{pmatrix} \sum_{i=1}^{p} w_{i1}x_i \\ \ldots \\ \sum_{i=1}^{p} w_{iq}x_i \end{pmatrix} = \sum_{i=1}^{p} \begin{pmatrix} w_{i1}x_i \\ \ldots \\ w_{iq}x_i \end{pmatrix} \qquad (6)$$

denotes the column vector of coordinates of $\hat{\mathbf{x}}[\mathbf{t}]$ in $\{\mathbf{w_k}\}_{1 \le k \le q}$ also referred to as the principal component scores.

Principal component aggregation The computation of the principal component scores z[t] can be performed within the network if each node i is initialized with the q scalar components $W_{i1};\ldots;W_{iq}$ of the principal component basis [14]. As proposed in [14], this initialization can done during a training stage where a set of measurements is first collected from the network and the related covariance matrix is estimated. This set of measurements is supposed to be large enough to provide an accurate estimate of the covariance matrix. Once the first q principal components are computed, they are sent back to the network and stored in the respective nodes (e.g., the i-th node stores only the elements $W_{i1};\ldots;W_{iq}$). While our proposition for distributing the computation of the principal components will be introduced in Section 3, here we limit to review the aggregation mechanism to compute the PC scores in a distributed manner.

The principal component scores can be computed by summing along the routing tree the vectors $(\mathbf{w}_{i1}x_i[t], .., \mathbf{w}_{iq}x_i[t])$ available at each node. The aggregation primitives take then the following form :

$$
\begin{aligned}
init(x_i[t]) &= \langle \mathbf{w}_{i1}x_i[t]; ..; \mathbf{w}_{iq}x_i[t] \rangle \\
f(\langle x_1; ..; x_q \rangle, \langle y_1; ..; y_q \rangle) &= \langle x_1 + y_1; ..; x_q + y_q \rangle
\end{aligned}
$$

where partial state records are vectors of size q.

APPLICATIONS

Once the transformed data z[t] reaches the base station, it can be used for multiple purposes and applications, such as approximate monitoring, feature extraction or event detection.

Approximate monitoring In approximate monitoring, the state of the process sensed by the WSN at time t is summarized by the vector

$$\begin{aligned}\hat{x}_t = e(z_1[t], ..., z_q[t]) &= (\hat{x}_1[t], ..., \hat{x}_n[t]) \\ &= W^T z[t]\end{aligned}$$

returned by the evaluator function at the base station.

If the number of components is p, the exact set of measurements collected by sensors is retrieved at the base station. Otherwise, for a number of components less than p, the vector $\hat{x}_t$ provides an optimal approximation to the real measurement vector in a mean square sense (Equation (4)). Note that the transformation in a principal component basis allows the design of simple policies to manage congestion issues, like discarding scores associated to the components with the lowest variance.

It is also interesting to note that a simple additional procedure can be set up to assess the accuracy of approximations against a user defined threshold. We know that the approximation $\hat{x}_i[t]$ for the i-th sensor $1 \leq i \leq p$ is given by (5) by:

$$\hat{x}_i[t] = \sum_{k=1}^{q} z_k[t] * w_{ik}$$

Since the elements $\{w_{ik}\}$ are already available at each node, by sending back to the routing tree the k principal component scores we can enable these two additional functionalities: (i) each sensor is able to compute the approximation retrieved at the sink, and (ii) each sensor is able to send a notification if the approximation error is greater than some user defined threshold. This scheme, referred to as supervised compression in [14], guarantees that all data eventually obtained at the sink are within ±€ of their actual measurements.

Dimensionality Reduction

An important class of WSN applications consists in inferring, on the basis of sensor measurements, some high level information such as the type or number of occurred events. For instance, a WSN monitoring the vibrations of a bridge could be used to answer queries concerning the type or number of vehicles that pass over the bridge. This class of problems is typically tackled by supervised learning techniques such as Naive Bayes classifiers, decision trees, or support vector machines [30, 31]. The PCA is in this context an effective preprocessing technique which simplifies the learning process by reducing the dimensionality of the problem [28, 31].

Event Detection

The use of PCA for identifying at the network scale unexpected events which are not detectable at the node level has been discussed in the literature [19, 20]. Detection of such events can be achieved by focusing on the low variance components. In fact, such components are expected to bear coordinates close to zero, as they typically account for sensor equipment noise. In this context, the principal component aggregation scheme can be used to deliver the value of low variance components, and the evaluator function takes the form of a statistical test checking if the coordinates on one or more low variance components is different from zero.

Tradeoffs

In this section, we will use the term default to denote the scheme that forwards all the measurements to the sink at each epoch, and the term PCAg to denote the scheme based on the principal component aggregation.

Let us first consider the distribution of the network load among sensor nodes, with a particular focus on its upper bound. The upper bound, henceforth referred to as highest network load, determines the network capacity, that is the amount of traffic that the network can handle at one time [6]. The highest network load is also related to the network lifetime in terms of time to first failure, as the radio is known to be the most energy consuming task of a wireless sensor [5, 25].

In the default scheme, the root of the routing tree is the node that sustains the highest network load as it forwards to the base station the measurements from all other sensors. Its radio throughput therefore causes the root node to have its energy exhausted first, which may lead the rest of the network to be disconnected from the base station.

In the PCAg scheme (Section 2.1.3), the network load is distributed in a more homogeneous manner.

The node $i_C^* = \arg\max_i C_i$ whose number of children is the highest becomes the limiting node in terms of highest network load. The network traffic at this node amounts to $q(C_{i_C^*}+1)$ packets per epoch, where q is the number of components retained. A PCAg scheme therefore reduces the highest network load in a configuration where

$$q(C_{i_C^*} + 1) \leq 2p - 1.$$

We illustrate the tradeoff by considering two extreme configurations. If only the first principal component is retained, the PCAg reduces the highest network load as $C_{i_C^*} < p$. On the contrary, if all the components are retained, i.e. $q = p$ the equation (7) simplifies into $pC_{i_C^*} \leq p - 1$ and the PCAg incurs a higher network load than the default scheme.

The main parameter of the approach is therefore the number of principal components to retain, which trades network load for sensing task accuracy. The amount of information retained by a set of q principal components depends on the degree of correlation among the data sources. Whenever a set of sensors collects correlated measurements, a small proportion of principal components is likely to support most of the variations observed by the network. The number of principal components to retain can be specified by the application, for example by means of a cross validation procedure on the accuracy of the sensing task at the base station [14]. It can be also be driven by network constraints or network congestion issues. In the latter case, a policy prioritizing packets containing the component scores of the principal components would allow to retain a maximum amount of information under network load constraints.

Finally, given a number of principal components, note that the highest network load may be further reduced if the algorithm that builds the routing tree uses $C_{i_c^*}$ as a parameter. The network load indeed depends on the number of children a node has. Although we do not address this aspect, routing strategies that aim at lowering $C_{i_c^*}$ can therefore further improve the efficiency of the PCAg scheme.

Computation of the Principal Components

Outline

This section addresses the initialization procedure whereby each node i can be initialized with the elements $W_{i1};\ldots;W_{iq}$ of the principal components $W_{i1};\ldots;W_{iq}$. In [14], we proposed a centralized approach for performing this initialization stage. This approach is first briefly recalled . We then extend this work by showing that a fully distributed implementation can perform this initialization process.

The proposed approach relies on the power iteration method (PIM), a classic iterative technique for computing the eigenvectors of a matrix [21], and on a simplifying assumption on the covariance structure which allows to estimate the covariance matrix in a distributed way. Section 3.3 presents and discusses the distribution of the covariance matrix computation and storage. We then show in Section 3.4 that by means of the distributed covariance matrix computation, the PIM can be implemented in an aggregation service in a distributed manner.

Each of these sections is concluded by a complexity analysis of the communication, computational, and memory costs. We finally discuss in Section 3.5 three methods for performing the initialization stage.

Centralized Approach

Let us suppose that t vectors of sensor measurements $\mathbf{x}[\tau] \in \mathbb{R}^p, \tau \in \{1, .., t\}$ are retrieved at the base station. Let $\mathbf{X}[t]$ denote the $p \times t$ matrix whose t -th column-vector is the vector $\mathbf{x}[\tau] \in \mathbb{R}^p$. An estimate of the covariance matrix $\mathbf{C}[t] = (c_{ij}[t])_{1 \le i,j \le p}$ is obtained at time t by computing

$$\begin{aligned} \mathrm{C}[t] &= \frac{1}{t}\sum_{\tau=1}^{t}(\mathrm{x}[\tau] - \bar{\mathrm{x}}[t])(\mathrm{x}[\tau] - \bar{\mathrm{x}}[t])^T \\ &= \frac{1}{t}\mathrm{X}[t]\mathrm{X}[t]^T - \bar{\mathrm{x}}[t]\bar{\mathrm{x}}[t]^T \end{aligned}$$

Where $\bar{\mathbf{x}}[t] = \frac{1}{t}\sum_{\tau=1}^{t}\mathbf{x}[\tau]$ is the average vector.

The covariance matrix can be updated recursively as new vectors of observations

x[t + 1] are made available at the base station. From Equation (8) it follows that

$$c_{ij}[t] = \frac{1}{t}S_{ij}[t] - \frac{1}{t^2}S_i[t]S_j[t] \tag{9}$$

where

$$S_i[t] = \sum_{\tau=1}^{t} x_i[\tau] = S_i[t-1] + x_i[t]$$
$$S_{ij}[t] = \sum_{\tau=1}^{t} x_i[\tau]x_j[\tau] = S_{ij}[t-1] + x_i[t]x_j[t] \quad (10)$$

This means that the storage of the quantities t, S_j and S_{ij} $1 \leq i, j \leq p$ in the base station makes possible the update of the covariance matrix estimation C[t] at time t. Once the covariance matrix is estimated, the principal components can be computed using a standard eigendecomposition method [21] and its elements communicated to the sensor nodes.

Scalability Analysis

Highest network load: The centralized estimation of the covariance matrix from a set of t vectors of observations requires t operations of type D as defined in Section 2.1.3. The root node supports the highest network load, which is therefore in $O(t_p)$. Once the eigenvector decomposition is performed at the base station, the root node then transmits the principal components to the sensor nodes, which requires qp operations of type F. Overall, the communication cost amounts to O(tp + qp).

Computational and memory costs: From Equation (10), the computational cost related to the covariance matrix at the base station is $O(tp^2)$. The memory cost for storing the matrix of observations and the covariances is $O(tp + p^2)$ or $O(p^2)$ if recursive updates are relied on. As far as the estimation of the principal components is concerned, the cost of standard eigendecomposition algorithm is $O(p^3)$ in terms of computation and $O(p^2)$ in terms of memory [21].

Distributed Estimation of the Covariance Matrix

This section presents an alternative methodology to distribute the computation and storage of the covariance matrix within the network. The methodology relies on the observation that sensors that are geographically distant often present low correlation values. This hypothesis often holds in real-world situations as shown in [32, 33]. This leads us to make the simplifying hypothesis that covariances between sensors which are out of reach is equal to zero. Note that the radio range can usually be tuned on wireless sensor platforms by increasing or reducing the power transmission level [25]. If available, this feature could be used to set the radio range such that nearby correlated sensors are within radio range.

Let Ni be the neighborhood of the sensor i, i.e., the set of sensors which are in the communication range of the i-th node. An instance of neighborhood is illustrated in Figure 1. Our simplifying hypothesis assumes that c_{ij} [t] = 0 for all j € Ni. This makes the computation of the covariance matrix scalable to large networks. The covariances between sensors i and j can indeed be computed recursively by Equation (10) once the sensor i keeps over time a record of t, Sj and Sij for j 2 Ni. This hypothesis will be referred to in the following as the local covariance hypothesis.

Positive Semi-definiteness Criterion

If the local covariance hypothesis is not verified, the distributed sparse matrix obtained by the local covariance hypothesis may lead to a globally non positive definite matrixz. The positive semi-definiteness is however a necessary (and sufficient) condition for a symmetric matrix to be a covariance matrix. To face this issue, the approach adopted in this article is to discard eigenvectors whose eigenvalues are found to be negative. This approach, according to [34], is the simplest way to transform a non positive definite matrix in a semi positive definite matrix. The detection of a negative eigenvalue will be detailed in Section 3.4.2, and used as a stopping criterion in the last line of Algorithm 3.4.2.

A good review of existing methods for transforming the approximated covariance matrix can be found in [34]. Our opinion is that it seems however difficult to implement these method in a distributed manner.

Note that despite its simplicity, the method that consists in discarding negative eigenvectors was observed by authors in [34] to provide good approximations to the covariance matrix.

Scalability Analysis

Highest network load: Each update requires a node i to send one packet (its measurement) and to receive $|\mathcal{N}_i|$ packets (other sensors' measurements). Let $i^*_{\mathcal{N}}$

N be the node that has the largest number of neighbors, i.e., $i^*_{\mathcal{N}} = \arg\max_i |\mathcal{N}_i|$

The highest network load is therefore $O(t|\mathcal{N}_{i^*_{\mathcal{N}}}|)$.

Computational and memory costs: From Equation (10), the updates of S_i and S_{ii} demand $3|\mathcal{N}_i|$ floating points additions/subtractions $2|\mathcal{N}_i|$ floating points multiplications and $2|\mathcal{N}_i|$ floating points divisions.

The highest computational cost therefore scales in $O(|\mathcal{N}_{i^*_{\mathcal{N}}}|)$. The number of variables to maintain at a sensor node is one integer and $2|\mathcal{N}_i|$ floating points numbers, so the highest memory cost amounts to $O(|\mathcal{N}_{i^*_{\mathcal{N}}}|)$.

Distributed Estimation of the Principal Components

The distribution of eigendecomposition algorithms is a fairly known research area [21]. However, the distribution sought in our problem should comply with the specificity that each node only has access to a single dimension of the problem and needs a specific entry of each principal eigenvector.

We propose in the following an approach that leverages the aggregation service to implement in a rather simple way the power iteration method (PIM), a well-known solution to the eigenvalue decomposition problem [21]. The resulting technique complies with the specific features mentioned above, and lets each node i locally identify the subset $\{w_{ik}\}$ required by the principal component aggregation.

After describing the PIM in Sections 3.4.1 and 3.4.2, we detail in Sections 3.4.3 and 3.4.4 its implementation in the aggregation service. The analysis of complexity of the method is provided in Section 3.4.5.

Power Iteration Method

Let C be a covariance matrix. The power iteration method starts with a random vector v_0. At the t-th iteration, the vector v_{t+1} is obtained by multiplying C by v_t, and by normalizing it. It can be easily shown [21] that v_t converges to the principal eigenvector w_1 of C, under the mild condition that v_0 is not strictly orthogonal to the principal eigenvector. The convergence rate is exponential, its base being the squared ratio of the two principal eigenvalues. The convergence criteria can be defined either as a minimum variation δ for v_t, or as a highest number of iterations t_{max} [21, 22]. Algorithm 3.4.1 outlines the different steps.

Standard power iteration algorithm $\mathbf{v}_0 \leftarrow$ arbitrary initialization $t \leftarrow 0$ $\mathbf{v}_t \leftarrow \mathbf{C}\mathbf{v}_t$ $\mathbf{v}_{t+1} \leftarrow \frac{\mathbf{v}_t}{\|\mathbf{v}_t\|}$

$t \leftarrow t+1$ $t > t_{max}$ and /or $\|\mathbf{v}_{t+1} - \mathbf{v}_t\| \leq \delta$

return v_t.

Note that as the method converges to the principal eigenvector w1, the normalizing factor $\|\mathbf{v}_t\|$ converges to the associated eigenvalue λ_1, as by definition :

$$\mathrm{C}\mathrm{w}_1 = \lambda_1 \mathrm{w}_1 \tag{11}$$

In practical settings, the power method quickly converges to a linear combination of eigenvectors whose eigenvalues are close, or to the eigenvector whose eigenvalue is the highest if eigenvalues are well separated. As our purpose here is to find the subspace that contains the signal, eigenvectors with close eigenvalues can be thought of as generating a subspace where similar amount of information are retained along any vector of the subspace. The convergence to a linear combination of eigenvectors with close eigenvalues is therefore deemed acceptable as far as principal component aggregation is concerned.

Computation of Subsequent Eigenvectors

The standard way to employ PIM in order to find the other eigenvectors (up to the number q) is the deflation method which consists in applying the PIM to the covariance matrix from which we have removed the contribution of the k principal eigenvector already computed [21]. We obtain

$$
\begin{aligned}
\mathbf{v}_{t+1} &= (\mathbf{C} - \sum_{l=1}^{k} \mathbf{w}_l \lambda_l \mathbf{w}_l^T)\mathbf{v}_t \\
&= \mathbf{C}\mathbf{v}_t - \sum_{l=1}^{k} \mathbf{w}_l \lambda_l \mathbf{w}_l^T \mathbf{v}_t
\end{aligned}
$$

which boils down to orthogonalize v_t with respect to all previously identified eigenvectors $\{\mathbf{w}_l\}_{1 \leq l < k+1}$.

A verification step must be added to the algorithm in order to detect negative eigenvalues. This can be achieved by comparing the sign of elements of v_t and v_{t+1} after convergence is assumed. From Equation (11), a negative eigenvalue implies that elements of v_t and v_{t+1} have pairwise different signs.

The criterion we define for determining the sign of an eigenvalue is

$$sign(\sum_{i=1}^{p} sign(v_t[i]v_{t+1}[i])$$

as taking the average sign of the sum of the signs over all pairwise products of v_t and v_{t+1} makes a more robust estimate of the sign of the eigenvalue in case of numerical errors. We rely on this additional criterion for stopping the computation of the principal components. The resulting process is given by Algorithm 3.4.2.

Modified power iteration algorithm for q eigenvectors $k \leftarrow 0$ $k \leftarrow k+1$ $t \leftarrow 0$ $\mathbf{v}_0 \leftarrow$ arbitrary initialization

$\mathbf{v}_t \leftarrow \mathbf{C}\mathbf{v}_t$ $\mathbf{v}_t \leftarrow \mathbf{v}_t - \sum_{l=1}^{k-1} \langle \mathbf{v}_t, \mathbf{w}_l \rangle \mathbf{w}_l$ $\mathbf{v}_{t+1} \leftarrow \frac{\mathbf{v}_t}{\|\mathbf{v}_t\|}$

$t \leftarrow t+1$ $t > t_{\max}$ or $\|\mathbf{v}_{t+1} - \mathbf{v}_t\| \leq \delta$ $\lambda_k \leftarrow \pm\|\mathbf{v}_t\|$ $\mathbf{w}_k \leftarrow \mathbf{v}_{t+1}$ $k = q$ or $\lambda_k < 0$

Implementation in the Aggregation Service

A fully distributed implementation of the algorithm is possible if the covariance matrix has been estimated as proposed in 3.3. The computation of one iteration therefore goes in two stages :

Computation of Cv

$$\forall i \in \{1, .., p\} , \ (\mathbf{Cv})[i] = \sum_{j=1}^{p} c_{ij} v[j]$$

As we made the assumption that:

$$\forall i \in \{1,..,p\}\ ,\ \forall j \notin \mathcal{N}_i\ ,\ c_{ij} = 0$$

The sum can be simplified :

$$\forall i \in \{1,..,p\}\ ,\ (\mathbf{Cv})\,[i] = \sum_{j \in \mathcal{N}_i} c_{ij} v[j]$$

Recall that each node i has available the set of covariances $(c_{ij}),\ j \in \mathcal{N}_i$

Therefore at each step t, node i only has to broadcast $v_t[i]$ and to receive the $v_t[j],\ j \in \mathcal{N}_i$ from its neighbors for updating locally its $\sum_{j \in \mathcal{N}_i} c_{ij} v[j] = (\mathbf{Cv})\,[i]$

The power iteration method can thus be implemented in a fully distributed manner.

Normalization and orthogonalization. The norm can be computed using the primitives detailed in Section 2.1.2, while the orthogonalization is obtained by means of these primitives:

$$\begin{aligned} init(v_t[i]) &= \langle (v_t[i]\mathbf{w}_\mathbf{l}[i])_{1 \le l \le k-1} \rangle \\ f(\langle X \rangle, \langle Y \rangle) &= \langle X + Y \rangle \\ e(\langle X \rangle) &= X. \end{aligned}$$

The resulting $k-1$ scalar products $\{\langle \mathbf{w}_l, \mathbf{v_t} \rangle\}_{1 \le l \le k-1}$ are then communicated to the sensors by means of an F operation, so that each sensor i can locally orthogonalize $\mathbf{v}_{t+1}[i] \leftarrow \mathbf{v}_t[i] - \sum_{l=1}^{k-1} \langle \mathbf{v}_t, \mathbf{w}_l \rangle \mathbf{w}_l[i]$.

Synchronization

The nodes have to be synchronized so that they all work on the same copy of v at the same time. The scheduling policy is presented in Algorithm 3.4.4. The implementation is straightforward since the aggregation service is synchronized.

Network synchronization scheme for the distributed power iteration algorithm $k \leftarrow 0 \; \forall i$ node i is initialized with $C[i, i]$. The nodes get their neighbor's value $(\mathbf{v_t}[j])_{j \in \mathcal{N}_i}$

The computation of C_{vt} is performed in parallel $\|\mathbf{v_t}\|$ and $\{\langle \mathbf{w}_l, \mathbf{v_t} \rangle\}_{1 \le l \le k-1}$ are computed by the aggregation service $\|\mathbf{v_t}\|$ and $\{\langle \mathbf{w}_l, \mathbf{v_t} \rangle\}_{1 \le l \le k-1}$ are fed back in the network For all i, $\mathbf{v_{t+1}}[i] \leftarrow \frac{(C\mathbf{v_t})[i] - \sum_{l=1}^{k-1} \langle \mathbf{v}_t, \mathbf{w}_l \rangle \mathbf{w}_l[i]}{\|\mathbf{v_t}\|}$ (in parallel on all nodes) convergence is obtained $\mathbf{w}_k \leftarrow \mathbf{v} \; k = q \text{ or } \lambda_k \le 0$.

Scalability Analysis

Highest network load: At each iteration, the computation of the product Cvt requires a node i to send one packet and to receive

$|\mathcal{N}_i|$ packets. The highest network load for this operation is therefore $O(|\mathcal{N}_{i^*_{\mathcal{N}}}|)$

The normalization step implies one A and one F operations (as defined in 2.1.3), and the orthogonalization step implies k 1 operations of type A and F respectively, where k is the index of the principal component computed. The highest network load for the computation of the q first principal components amounts to $O(q|\mathcal{N}_{i^*_{\mathcal{N}}}| + q^2|\mathcal{C}_{i^*_{\mathcal{C}}}|)$

The highest network load related to the power iteration method is therefore quadratic in the number of principal components computed. Depending on the constraints of the network, it may be more interesting to retrieve the approximated covariance matrix from the network at the base station for a centralized computation of the eigenvectors. The tradeoff depends here on the communication and computational resources available in the network, and will be discussed more in detail in Section 3.5.

Computational and memory costs For a node i, the cost of the computation of $\mathbf{Cv}_t$ is $O(|\mathcal{N}_i|)$ The cost of the normalization step is $O(|\mathcal{C}_i|)$ for node i, and the orthogonalization step is $O(k|\mathcal{C}_i|)$

The overall highest computational cost therefore amounts to $O(q(|\mathcal{N}_{i^*_{\mathcal{N}}}| + |\mathcal{C}_{i^*_{\mathcal{C}}}|)$ Regarding memory costs, each node i needs to maintain variables for storing its local w_{ik} and its neighbors parameters $w_{jk}, j \in \mathcal{N}_i$.

The complexity of the highest memory cost is therefore $O(q + |\mathcal{N}_{i^*_{\mathcal{N}}}|)$.

SUMMARY

The benefits of the different approaches for computing the covariance matrix and its principal components depend on the communication, computational and memory constraints available in the network nodes and at the base station. A summary of the complexities of these approaches is given in Table 1.

Table 1. Scalability of the centralized, hybrid and distributed schemes.

Operation	Communication	Computation	Memory
Covariance			
Centralized	$O(pT)$	$O(p^2T)$	$O(p^2)$
Distributed	$O(\lvert\mathcal{N}_{i^*_N}\rvert T)$	$O(\lvert\mathcal{N}_{i^*_N}\rvert T)$	$O(\lvert\mathcal{N}_{i^*_N}\rvert)$
Eigenvectors			
Centralized	$O(qp)$	$O(p^3)$	$O(p^2)$
Distributed	$O(q^2\lvert\mathcal{N}_{i^*_N}\rvert)$	$O(q(\lvert\mathcal{N}_{i^*_N}\rvert + \lvert\mathcal{C}_{i^*_C}\rvert))$	$O(q + \lvert\mathcal{N}_{i^*_N}\rvert)$

EXPERIMENTAL RESULTS

This section illustrates by means of a compression task the ability of the proposed approach to properly identify the principal component subspace, and discusses the tradeoffs involved between the compression accuracy and the communication costs. The experimental study is based on a set of real world temperature measurements involving a network of 52 sensors (Section 4.1), and on a network simulation used to vary the structure of the routing trees to study the impact of the network topology on the network load (Section 4.2). Results on the ability of a few components to retain most of the information are first illustrated in Section 4.3, followed in Section 4.4 by an analysis of the network loads as the number of retained components increases. The ability of the distributed PCA to properly identify the principal components is then studied in Sections 4.5 and 4.6, which focus on the covariance matrix approximation and eigenvector extraction stages, respectively. Section 4.7 concludes by a discussion on the main results.

Dataset description

Experiments were carried out using a set of five days of temperature readings obtained from a 54 Mica2Dot sensor deployment at the Intel research laboratory at Berkeley [35]. The sensors 5 and 15 were removed as they did not provide any measurement. The readings were originally sampled every thirty-one seconds. A preprocessing stage where data was discretized in thirty second intervals was applied to the dataset. After preprocessing, the dataset contained a trace of 14400 readings from 52 different sensors. The code associated to the preprocessing and the network simulation was developed in R, an open source statistical language, and is available from the authors' web site [36].

Examples of temperature profiles and dependencies between measurements are reported in Fig. 4 and 5, respectively. The sensors 21 and 49 were the least correlated ones over that time period, with a correlation coefficient of 0:59. They were situated on opposite sides of the laboratory. Temperature over the whole set of data ranged from about 15^{0}C to 35^{0}C.

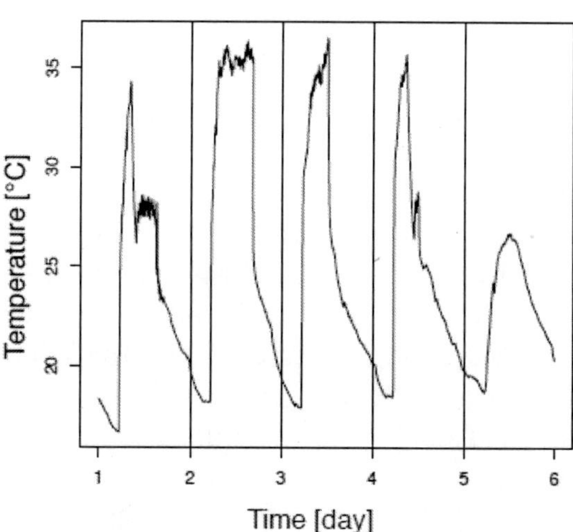

Figure 4. Temperature measurements collected by sensor 21 over a five day period.

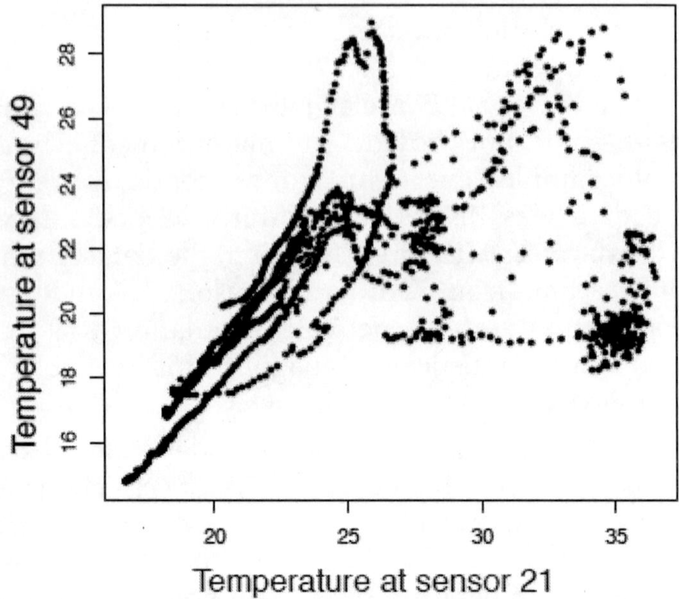

Figure 5. Examples of the dependencies between the measurements of sensor 21 and sensor 49.

Network Simulation

The positions of the sensors are provided in [35], and the distribution of the sensors in the laboratory can be seen in Fig. 6. We analyzed the communication costs in different routing trees which were generated in the following way. The root node was always assumed to be the top right sensor node in Fig. 6 (node 16 in [35]). The routing trees were generated on the basis of the sensor positions and the radio range was varied from 6 meters (minimum threshold such that all sensor could find a parent) to 50 meters (all sensors in radio range of the root node). Starting from the root node, sensors were assigned to their parent in the routing tree using a shortest path metric, until all sensors were connected. An illustration of the routing tree obtained for a maximum communication range of 10m is reported in Fig.6.

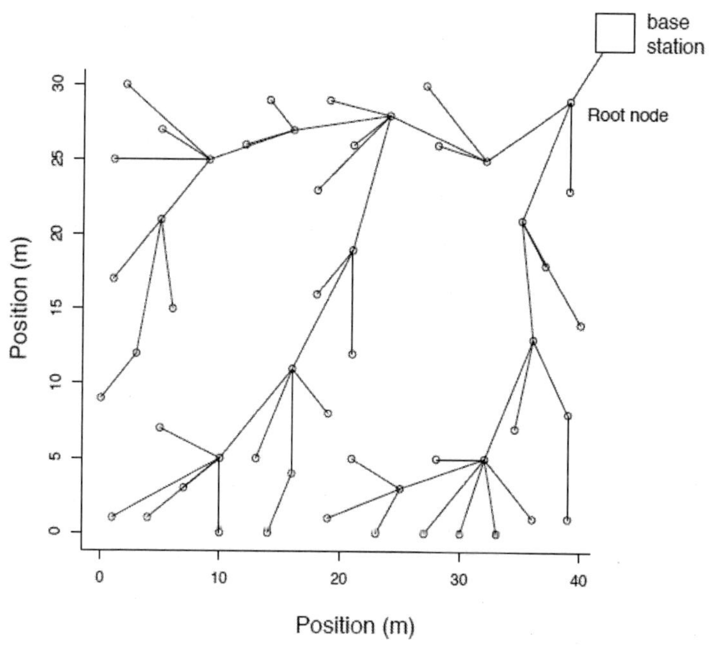

Figure 6. Map of sensors and the routing obtained for a connectivity of 10 meters. The root node connecting the sink is the top right sensor. The routing tree has a depth of seven.

Principal Component Aggregation

We study in this section the amount of variance that the PCA can retain in the measurements for different training periods and number of components. We rely on a 10-fold cross validation technique to simulate the fact that the measurements used to compute the principal components are not used for assessing the accuracy of the approach. More precisely, the dataset is split in ten consecutive blocks (1440 observations – i.e., half a day of measurements). Each of the ten blocks is used in turn as a training set to compute the covariance matrix and its eigenvectors, while the remaining observations, referred to as test set, are used to estimate the percentage of retained variance. This provides ten estimates of the percentage of retained variance on measurements not used for computing the eigenvectors, which we average to obtain an overall estimate of the method.

Fig. 7 provides the average retained variance on the 10 test sets for the first 25 principal components. The upper line gives the average amount of variance retained when the principal components are computed with the test sets, and provides an upper bound on the compression efficiency that the PCA can achieve on this dataset. The lower curve gives the average amount of variance retained on the test set when the components are computed with the training set. This figure shows that the first principal component accounts on average for almost 80% of the variance, while 90% and 95% of variance are retained with 4 and 10 components, respectively. The confidence level of these estimates was about _5%. Additional experiments were run using K-fold cross validation with K ranging from 2 to 30. The percentages of retained variance on the test data blocks tended to decrease with K. Losses of a few percents were observed for K higher than 15 (less than nine hours of data).

The amount of retained variance increases very fast with the first principal component, and becomes almost linear after about ten components. A linear increase of retained variance with the number of principal components reflects the fact that the components obtained by the PCA are actually no better than random components [28]. From Fig. 7, it therefore seems that from 10 or 15 components onwards, the remaining variations can be considered as white noise.

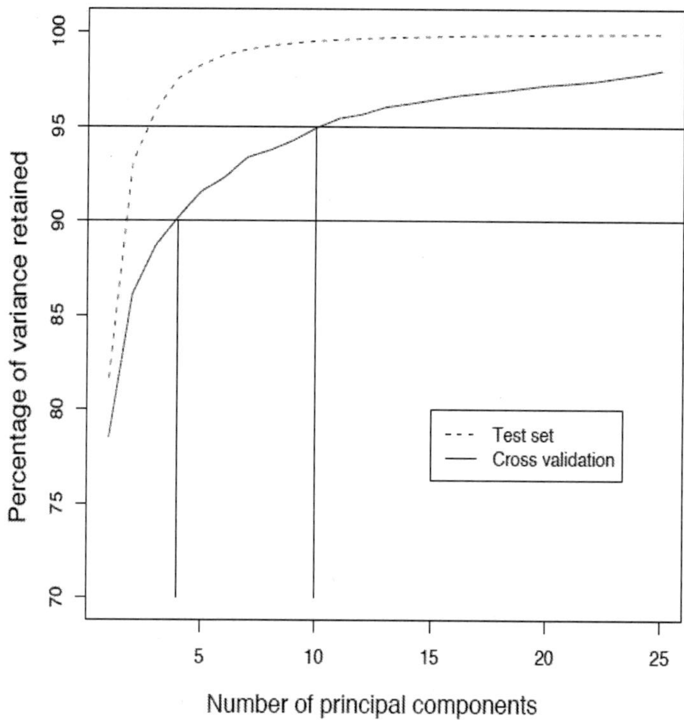

Figure 7. Capacity of principal components to retain the measurement variance.

Fig. 8 illustrates the approximations obtained during the first round of the cross validation (i.e., principal components are computed from the first 12 hours of measurements) for the sensor 49, using one, five and ten principal components. A single principal component provides rough approximations, which cannot account for fine-grained details of some of the sensor measurements. For example, the stabilization of the temperature around 20_C around noon during the second, third and fourth day (probably due to the activation of an air conditioning system at a location close to sensor 49) are not captured by the approximations. Increasing the number of principal components allows to better approximate the local variations, and passing to five components provides for example a much better approximation of the sensor measurements.

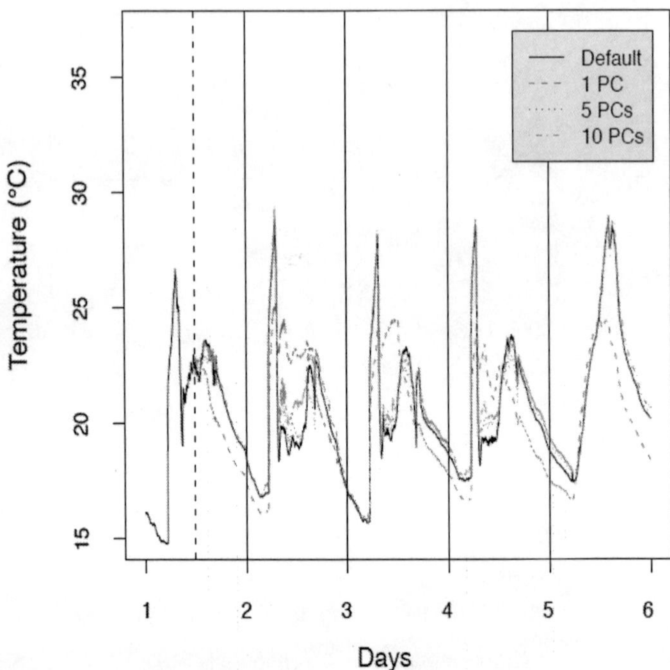

Figure 8. Approximations obtained on the test set for the sensor 21, using one, five and fifteen principal components. The covariance matrix was computed from the first twelve hours of measurements (before the vertical dashed line).

Communication Costs

We compare in the following the communication costs entailed by the default and PCAg schemes for different types of routing trees. Fig. 9 reports the total number of packets processed by nodes (received and transmitted) during an epoch, for the default and PCAg scheme, as the radio communication range increases. The subfigure on the left reports the overall sensor network load, while subfigures in the middle and on the right detail with boxplots the distributions of the network load per node in the network, for the default and PCAg schemes, respectively.

Given that sensors choose as their parent the sensor within radio range that is the closest to the base station, increasing the radio communication range typically leads the routing tree to have a smaller depth, and its nodes to have a higher number of children. As was pointed out in Section 2.5, this is detrimental to the PCAg scheme, as this may lead the numbers of packets received by a node to be higher than in the default scheme.

For the default scheme, increasing the radio range reduces the overall sensor network load (Fig. 9, left), and eventually leads all the nodes but the root to only transmit one packet (Fig. 9, middle, radio range of 50m). In this extreme case, the routing tree has depth one, and all nodes but the root are leaf nodes. Note that the highest network load does not depend on the tree topology. This highest load is sustained by the root node which, whatever the depth of the tree, is required to forward other node's packets (i.e., 51 measurements to receive and send), and to send its own measurement to send. Its network load is therefore of 103 packets per epoch.

For the PCAg scheme, we first report results for the extraction of one component. A nice feature of aggregation is that the overall network load (Fig. 9, left) does not depend on the topology, thanks to the fact that forwarding does not increase the number of transmissions. Looking at the details of the distribution of the network load per node (Fig. 9, right), it is interesting to see that increasing the radio range has a reverse effect for the PCAg scheme, as it tends to increase the network load. This is a direct consequence of the increased number of children induced by routing trees with smaller depths. Eventually, for a fully interconnected network, we observe the same effect than for the default scheme, where all the sensors send only one packet, while the root node sustains the higher network load due to the forwarding task. Note however than thanks to the aggregation process, it only sends one packet, which reduces to 52 packets per epoch its network load (one packet transmitted and 51 packets received). The extraction of one component therefore decreases the network load supported by the root node. The network load incurred by k components is obtained by multiplying the values found in Fig. 9 by k.

The PCAg scheme may therefore be less efficient than the default scheme if many components are extracted.

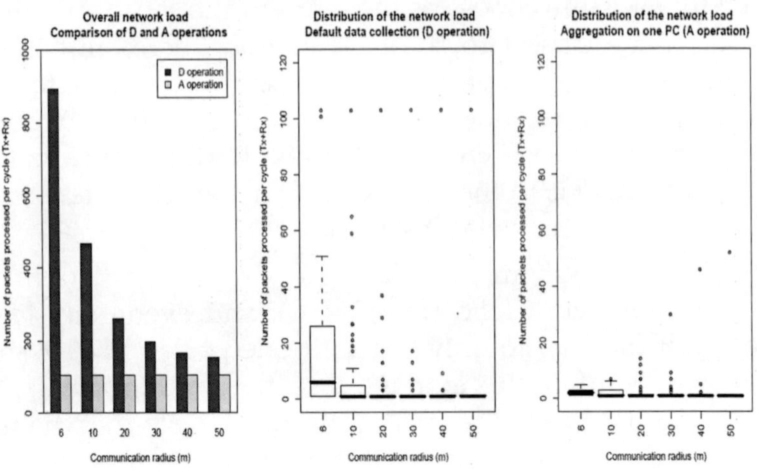

Figure 9. Communication costs entailed by D and A operations. The overall network load (left) is in all cases higher for a D operation than for an A operation. Details of the distribution of the network load per node reported as boxplots (middle and right) reveal that the distributions of the load are more balanced using aggregation. Aggregation also leads to reduce the highest network load. This is illustrated in Fig. 10 where the number of packets processed (received and sent) is reported as a function of the number of principal components extracted for a radio range of 10 (Illustrated in Fig. 6). In this routing tree, the highest number of children is 6. For the extraction of one PC, the highest network load is therefore of 7, i.e., 6 receptions and one transmission, to be compared with the highest network load of 103 for the root node in the default scheme. This results in a reduction of about 85% of the network load. Extracting more than 15 components leads however the highest network load to be higher than in the default scheme, as the sensor node aggregating the packets from its 6 children will sustain a network load of 105 packets per epoch. The overall network load generated by the aggregation of k components is in the same manner k times the load generated by of the aggregation of one component. For a communication radius of 10 meters, this overall load was of 103 packets as reported in Fig. 9, left, to be compared with an overall load of 466 packets for the default scheme. The overall load generated by the aggregation process is therefore higher than the load of the default scheme from 5 components. This is due to the fact that aggregation, while decreasing the load of the most solicited node, increases evenly the load of all other nodes.

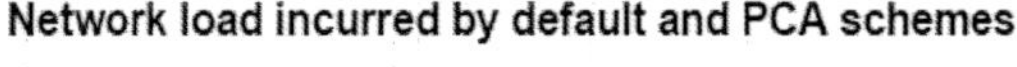

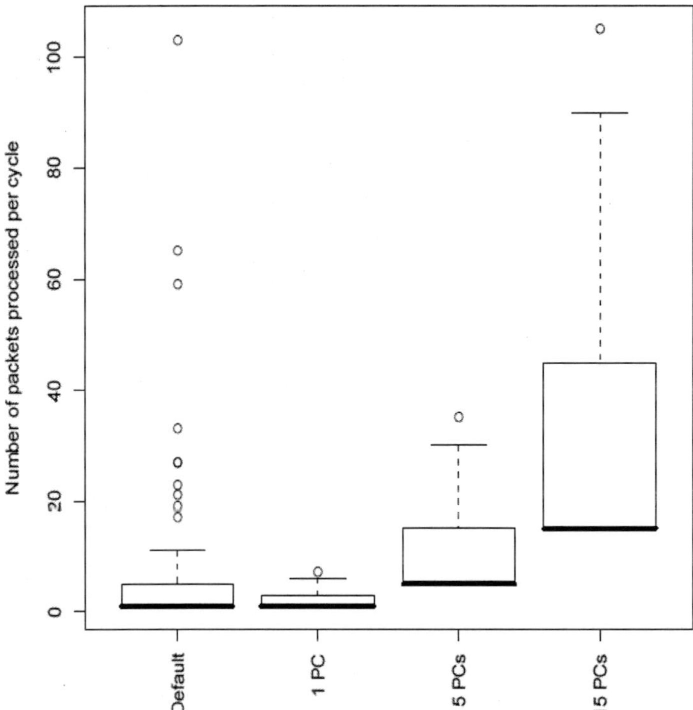

Figure 10. Comparison of the communication costs entailed by a D operation, and A operations with 1, 5 and 15 principal components. Radio range is 10 meters.

Distributed Covariance Matrix

We study in this section the ability of the local covariance hypothesis to properly identify the principal component subspace, and discuss how the accuracy loss is counterbalanced by gains in energy consumption and network load. In Fig. 11, the upper curve gives the amount of variance retained if all covariances are computed, and is the same as in Fig. 7. Lower curves correspond to the percent age of variance retained as the radio range of sensors is decreased, and illustrate the fact that the local covariance hypothesis may have as a negative consequence a loss of accuracy for identifying the principal components. If the local covariance hypothesis does not hold, the loss can be high, as is illustrated by a radio range of 6 meters. This loss is however attenuated when the number of principal components increases. In any case, the subspace obtained by computing the principal components from the approximate covariance matrix is significantly better in retaining information than a random subspace.

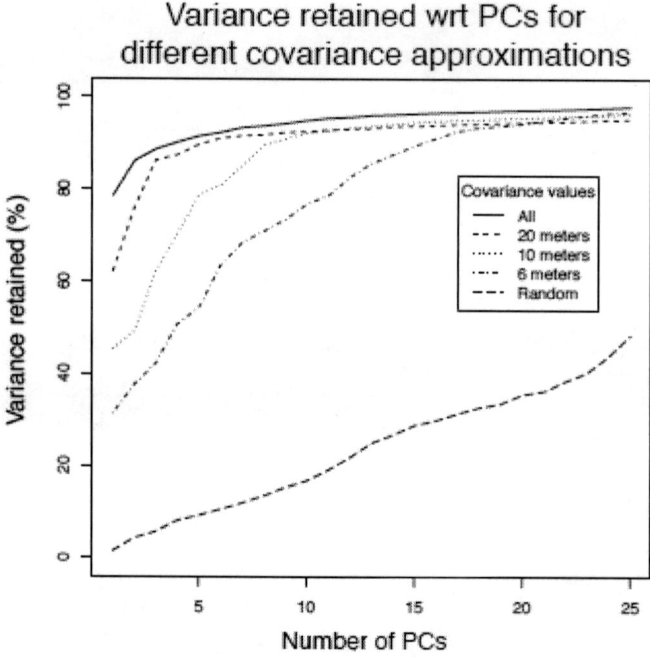

Figure 11. Capacity of the principal components to retain the measurement variance.

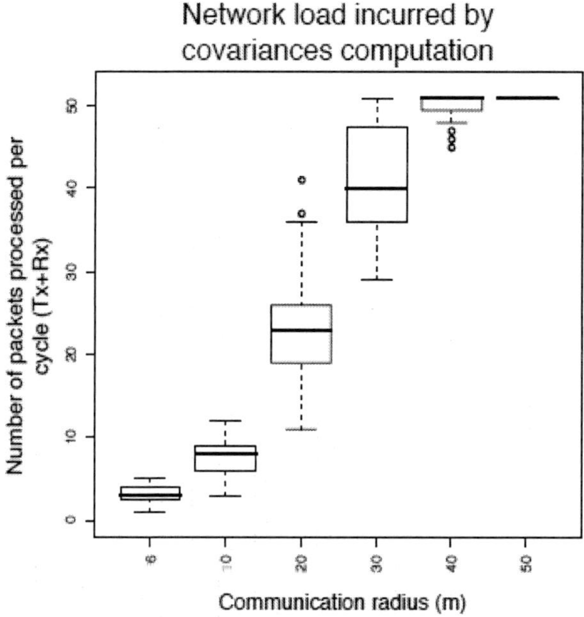

Figure 12. Network load incurred by local covariances updates.

We note that reducing the radio range also decreases the energy spent in communication, as a lower transmission power leads to a lower energy consumption for the radio. We illustrate in Fig. 12 the potential savings in terms of network load obtained by decreasing the radio range. This figure reports the distribution of the network loads related to the computation of the covariance matrix for varying radio ranges. The average network load increases with the radio range, as the neighborhood of sensors gets larger. Fig. 11 and 12 provide together an illustration of the tradeoff between the accuracy and the communication costs entailed by the local covariance hypothesis. A radio range range larger than 20 meters does not bring much gains in terms of accuracy, while it strongly increases the network load.

Also, it is interesting to compare the results reported in Fig. 12 and Fig. 9. The highest network load caused by an update is in all cases lower using the distributed covariance matrix scheme (52 packets processed against 101 at the root for the default data gathering operation). The average network load can however be higher as the radio range increases, due to the fact that nodes close to the leaves process more packets in the distributed scheme than in the centralized scheme.

Distributed Principal Component Computation

We finally discuss the ability of the power iteration method to properly identify the principal components. The main parameter for the estimation of the eigenvectors is the convergence criteria used. We illustrate in Fig. 13 the difference in accuracy obtained on the test set, for different convergence criteria, between the set of exact eigenvectors (Computed in a centralized manner with the QR method), and the set of approximated eigenvectors obtained by means of the power iteration method. The convergence threshold _ was set to 10□3, and we tested the accuracy obtained for 5, 10, 20, 30, 40 and 50 iterations.

Results reported are averages of ten-fold cross validations, and the confidence level of the results was about 2%.

This Figure allows to experimentally illustrate a set of possible behaviors, which are understood by keeping in mind that the rate of convergence of the power method depends on the ratio of the two dominant eigenvalues. Typically, in correlated data, the ratio of subsequent eigenvalues decreases exponentially (which can be seen qualitatively in Fig. 7), making the convergence speed lower as the number of principal components computed increases. As a result, few iterations are usually enough to converge to the first eigenvectors. This is seen in Fig. 13 for the first PC, for which as little as five iterations are enough to nicely converge. In the computation of the subsequent PCs, we observe that the number of iterations required to properly converge is higher (about 20 iterations led to accuracy similar to the centralized approach). If the number of iterations is not high enough, the power iteration leads to a PC that may not represent the data as well as the centralized approach, as is observed from the second PC for a maximum number of five iterations. Note that, given the fact that PCs are estimated on the basis of past measurements whose distribution is not exactly the same as upcoming measurements, it may also happen that approximations provide better accuracies than the centralized approach. This explains the gains in accuracy obtained around the 8-th component where the difference between subsequent eigenvalues gets very low.

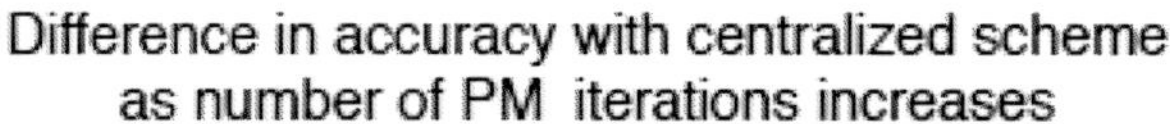

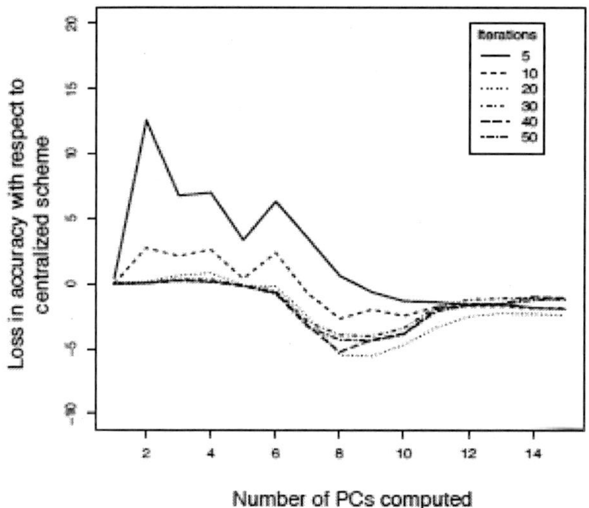

Figure 13. Comparison of the accuracy obtained using the exact and the approximated eigenvectors, for different number of iterations and principal components.

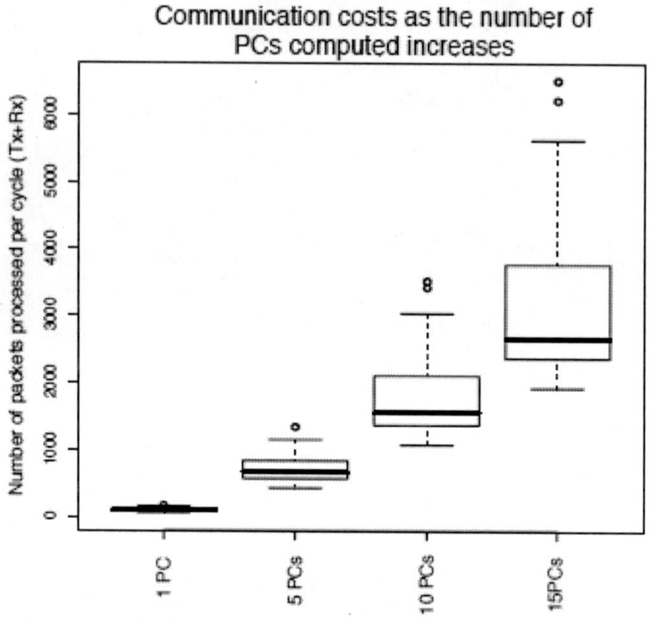

Figure 14. The network load is quadratic in the number of principal components. Radio range is 10 meters.

Additional experiments were also run to observe the consequence of the local covariance hypothesis on the non positive semi-definiteness of the matrix. We observed that negative eigenvalues could lead the algorithm to stop at stages as early as between 5 to 10 components (this was observed for radio ranges of 30 and 40 meters). It is however important to note that despite the early stopping of the algorithm, the

set of principal components identified was enough to retain more than 90% of the variance (cf. Fig. 11).

Finally, we report in Fig. 14 estimates of the communication costs entailed by the computation of the principal components. The radio range was of 10 meters, and the costs were computed following the analysis of Sections 2.1.3 and 3.5. While about two hundred packets per node suffice for the computation of the first eigenvector, this quantity can reach a significant network load of 6000 packets on average for the computation of 15 components. As discussed in

Section 3.5, the main advantage of the distributed computation of the principal components lies in the distribution of the computational and memory costs, and the approach does not scale well to the computation of a large number of principal components.

SUMMARY

The main parameter of the principal component aggregation, which determines the tradeoff between the compression accuracy and the communication costs, is the number of principal components to extract. As illustrated in Fig. 10 or 14, the network load incurred by the computation of the principal components quickly increases with their number. The PCAg scheme therefore proves useful only if a few components are required. In such a case, we emphasize that the gains can be dramatic, as was reported in Fig. 10 or 14. The amount of information retained by a set of principal components depends on the correlation existing between sensor measurements. We illustrated the method for a compression task using a realworld temperature dataset where measurements had an degree of correlation that can be assumed to be representative of a typical sensor network scenario. An interesting result to recall is that for 5 principal components, 90% of the variance could be retained. This was shown to return a nice approximation of the original signals (Fig. 8), and to reduce the highest network load from about 100 packets per epoch to about 40 packets per epoch (Fig. 10).

Regarding the scalability, an interesting property of the PCAg is that the network load is better distributed in the network. This property first prevents congestion issues observed at the root node using the default scheme. Second, as the radio communication is a primary source of energy depletion for a wireless node, it better distributes the lifetime of the nodes in the network. The PCAg therefore provides an interesting framework for dealing with data in multi-hop networks.

Related Work

The distribution of the PCA has been addressed in the signal processing and information theory communities [4, 37–39]. These approaches, based on distributed source coding, involve compression techniques related to Slepian-Wolf and Wyner-Ziv coding together with Karhunen-Loeve transform for compressing data in distributed systems. Each component of the system is however assumed to observe several dimensions of the problem, and they require the base station to support the task of defining the coding scheme. The network architectures and communication costs analyses are also left as open research areas. Although these approaches may lead to attractive applications for WSN, they still remain in that respect at an early research stage. In the fields of machine learning and data mining, distributed PCA schemes have also been addressed in [40, 41] and [19]. The former approaches aimed at combining the PCA results obtained from different sources that share the same variables (vertically distributed databases). The latter work provided an architecture aimed at network anomaly detection, where the PCA is first computed in a centralized manner, and a subsequent distributed and adaptive feedback mechanism is proposed. Despite the similarity in the choice for the method name, the corresponding work is however clearly different from the approach proposed in this article.

The distributed computation of eigenvectors of large and sparse matrices has been tackled from several angles in the literature, and good reviews of existing state of the art techniques are for example detailed in [21, 22]. The approach proposed in this article is however assumed to be innovative to the best of authors' knowledge, as it leverages two specific architectural properties of sensor networks. First, a wireless sensor network is an intrinsically distributed system where each component only captures one dimension of the system variations. Second, the communication constraints can be coupled with the local covariance hypothesis.

Recent work in the domain of link analysis and recommendation systems has led authors in [42] to propose a distributed implementation of the power iteration method, which closely matches the approach proposed in this article. Their algorithm aims at computing the principal eigenvector of the adjacency matrix of a peer-to-peer network, which leads to the ranking of the network components, in a way similar to the page rank algorithm. The underlying hypotheses of the distributed system are interestingly close to ours (each component has only access to one dimension the problem, and can communicate with components whose data is related). The network structure is however different, as no coordinating entity is assumed. This leads authors to rely on the harmonic mean to achieve the normalization step of the power iteration method. The computation of subsequent eigenvectors is also not addressed.

Related work on the use of basis change for sensor networks has been addressed in [43], where authors proposed the radical position of using random bases to project sensor measurements. The work is analyzed in the context of compressed sensing, an emerging field in signal processing (see also [44]). Their work has however mainly focused on the theoretical ability of random bases to retain the sensor measurements variations.

CONCLUSION

This article extended previous work related to the distribution of the principal component analysis by presenting an implementation suitable for wireless sensor networks. The approach relies on the hypothesis that sensor measurements collected by distant sensors are uncorrelated, and was shown to provide significant gains in terms of radio communication. Additionally, we showed that the distributed principal component analysis led to balance the network load among the sensors, making the method particularly suitable for multi-hop sensor networks. We plan to extend this work by showing that spatiotemporal aggregation and independent component analysis can also be formulated in the same framework.

ACKNOWLEDGMENTS

This work was supported by the COMP2SYS project, sponsored by the Human Resources and Mobility program of the European Community (MEST-CT-2004-505079), and by the PIMAN project, supported by the Institute for the Encouragement of Scientific Research and Innovation of Brussels, Belgium.

REFERENCES

1. Kumar, S.; Zhao, F.; Shepherd, D. Collaborative signal and information processing in microsensor networks. Signal Processing Magazine, IEEE 2002, 19(2), 13–14.
2. Intanagonwiwat, C.; Govindan, R.; Estrin, D. Directed diffusion: A scalable and robust communication paradigm for sensor networks. In Proceedings of the ACM/IEEE International Conference on Mobile Computing and Networking, pages 56–67, 2000.
3. Pattem, S.; Krishnamachari, B.; Govindan, R. The impact of spatial correlation on routing with compression in wireless sensor networks. In Proceedings of the third international symposium on Information processing in sensor networks, pages 28–35. ACM Press, 2004.
4. Xiong, Z.; Liveris, A.; Cheng, S. Distributed source coding for sensor networks. Signal Processing Magazine 2004, 21(5), 80–94.
5. Akyildiz, I.; Su, W.; Sankarasubramaniam, Y.; Cayirci, E. Wireless sensor networks: a survey. Computer Networks 2002, 38(4), 393–422.
6. Ilyas, M.; Mahgoub, I.; Kelly, L. Handbook of Sensor Networks: Compact Wireless and Wired Sensing Systems. CRC Press, 2004.
7. Krishnamachari, L.; Estrin, D.; Wicker, S. The impact of data aggregation in wireless sensor networks. In Proceedings of the Distributed Computing Systems Workshops, pages 575–578, 2002.
8. Heidemann, J.; Silva, F.; Intanagonwiwat, C.; Govindan, R.; Estrin, D.; Ganesan, D. Building efficient wireless sensor networks with low-level naming. In Proceedings of the eighteenth ACM symposium on Operating systems principles, pages 146–159. ACM Press, 2001.
9. Intanagonwiwat, C.; Estrin, D.; Govindan, R.; Heidemann, J. Impact of network density on data aggregation in wireless sensor networks. In Proceedings of the 22nd International Conference on Distributed Computing Systems., pages 457–458, 2002.
10. Madden, S.; Franklin, M.; Hellerstein, J.; Hong, W. TAG: a Tiny AGgregation Service for Ad-Hoc Sensor Networks. In Proceedings of the 5th ACM Symposium on Operating System Design and Implementation (OSDI), volume 36, pages 131 - 146. ACM Press, 2002.
11. Madden, S.; Franklin, M.; Hellerstein, J.; Hong, W. TinyDB: an acquisitional query processing system for sensor networks. ACM Transactions on Database Systems 2005, 30(1), 122–173.
12. Yao, Y.; Gehrke, J. The cougar approach to in-network query processing in sensor networks. ACM SIGMOD Record 2002, 31(3), 9–18.
13. Burri, N.; Wattenhofer, R. Dozer: ultra-low power data gathering in sensor networks. In Pro ceedings of the 6th international conference on Information processing in sensor networks, pages 450–459. ACM Press, 2007.

14. Le Borgne, Y.; Bontempi, G. Unsupervised and supervised compression with principal component analysis in wireless sensor networks. In Proceedings of the Workshop on Knowledge Discovery from Data, 13th ACM International Conference on Knowledge Discovery and Data Mining, pages 94–103. ACM Press, 2007.

15. . Hyvarinen, A.; Karhunen, J.; Oja, E. Independent Component Analysis. J. Wiley New York, 2001.

16. Li, J.; Zhang, Y. Interactive sensor network data retrieval and management using principal components analysis transform. Smart Materials and Structures December 2006, 15, 1747–1757(11).

17. Bontempi, G.; Le Borgne., Y. An adaptive modular approach to the mining of sensor network data. In Proceedings of the Workshop on Data Mining in Sensor Networks, SIAM SDM, pages 3–9. SIAM Press, 2005.

18. Duarte, M.; Hen Hu, Y. Vehicle classification in distributed sensor networks. Journal of Parallel and Distributed Computing 2004, 64(7), 826–838.

19. Huang, L.; Nguyen, X.; Garofalakis, M.; Jordan, M.; Joseph, A.; Taft, N. In-network PCA and anomaly detection. In B. Scholkopf, J. P.; Hoffman, T., editors, Proceedings of the 19th conference on Advances in Neural Information Processing Systems. MIT Press, 2006.

20. Lakhina, A.; Crovella, M.; Diot, C. Diagnosing network-wide traffic anomalies. In Proceedings of the 2004 conference on Applications, technologies, architectures, and protocols for computer communications, pages 219–230. ACM Press, 2004.

21. Bai, Z.; others. Templates for the Solution of Algebraic Eigenvalue Problems: A Practical Guide. Society for Industrial and Applied Mathematics, 2000.

22. Golub, G.; Van Loan, C. Matrix computations. Johns Hopkins University Press, 1996.

23. Akkaya, K.; Younis, M. A survey on routing protocols for wireless sensor networks. Ad Hoc Networks 2005, 3(3), 325–349.

24. TinyOS. Project Website: http://www.tinyos.net.

25. Polastre, J.; Szewczyk, R.; Culler, D. Telos: enabling ultra-low power wireless research. In Proceedings of the 4th International Symposium on Information Processing in Sensor Networks, pages 364–369, 2005.

26. Zhao, F.; Guibas, L. Wireless Sensor Networks: An Information Processing Approach. Morgan Kaufmann, 2004.

27. Raghunathan, V.; Srivastava, C. Energy-AwareWireless Microsensor Networks. Signal Processing Magazine 2002, 19(2), 40–50.

28. Jolliffe, I. Principal Component Analysis. Springer, 2002.

29. Miranda, A. A.; Le Borgne, Y.-A.; Bontempi, G. New routes from minimal approximation error to principal components. Accepted for publication in the Neural Processing Letters 2008.

30. Duda, R.; Hart, P.; Stork, D. Pattern Classification. Wiley-Interscience, 2000.

31. Webb, A. Statistical Pattern Recognition. Hodder Arnold Publication, 1999.
32. Barrenetxea, G. SensorScope: on-line urban environmental monitoring network. Geophysical Research Abstracts 2007, 9, 07501.
33. Jindal, A.; Psounis, K. Modeling spatially-correlated sensor network data. Sensor and Ad Hoc Communications and Networks, 2004. IEEE SECON 2004. 2004 First Annual IEEE Communications Society Conference on 2004, pages 162–171.
34. Rousseeuw, P.; Molenberghs, G. Transformation of non positive semidefinite correlation matrices. Communications in Statistics-Theory and Methods 1993, 22(4), 965–984.
35. Intel Lab Data webpage. http://db.csail.mit.edu/labdata/labdata.html.
36. Wireless Sensor Lab. Machine Learning Group. University of Brussels. http://www.ulb. ac.be/di/labo/index.html.
37. Vosoughi, A.; Scaglione, A. Precoding and Decoding Paradigms for Distributed Data Compression. IEEE Transactions on Signal Processing 2007.
38. Pradhan, S.; Ramchandran, K. Distributed source coding using syndromes (DISCUS): design and construction. IEEE Transactions on Information Theory 2003, 49(3), 626–643.
39. Gastpar, M.; Dragotti, P. L.; Vetterli, M. The Distributed Karhunen-Lo`eve Transform. IEEE Transactions on Information Theory 2006, 52(12), 5177–5196.
40. Bai, Z.-J.; Chan, R.; Luk, F. Principal component analysis for distributed data sets with updating. In Advanced Parallel Processing Technologies, volume 3756, pages 471–483. Springer Berlin, Heidelberg, 2005.
41. Kargupta, H.; Huang, W.; Sivakumar, K.; Park, B.; Wang, S. Collective Principal Component Analysis from Distributed, Heterogeneous Data. In Proceedings of the 4th European Conference on Principles of Data Mining and Knowledge Discovery, pages 452–457. Springer-Verlag, 2000.
42. Jelasity, M.; Canright, G.; Engø-Monsen, K. Asynchronous Distributed Power Iteration with Gossip-Based Normalization. In Proceedings of Euro-Par, volume 4641 of Lecture Notes in Computer Science, pages 514–525. Springer, 2007.
43. Duarte, M. F.; Sarvotham, S.; Baron, D.; Wakin, M. B.; Baraniuk, R. G. Distributed compressed sensing of jointly sparse signals. In Proceedings of the 39th Asilomar Conference on Signals, Systems and Computation, pages 1537–1541, 2005.
44. Donoho, D. Compressed Sensing. IEEE Transactions on Information Theory 2006, 52(4), 1289–1306.

Chapter 6

ESTIMATION OF ATMOSPHERIC PATH DELAYS IN TERRASAR-X DATA USING MODELS VS. MEASUREMENTS

Michael Jehle [1;*], Donat Perler [2], David Small[1]
Adrian Schubert [1] and Erich Meier [1]

[1] Department of Geography, University of Zurich, Winterthurerstrasse 190, CH-8057 Zurich, Switzerland

[2] Institute of Geodesy and Photogrammetry, ETH Zurich, HPV G56, Schafmattstrasse 34, CH-8093 Zurich, Switzerland

E-mails: michael.jehle@geo.uzh.ch; donat.perler@geod.baug.ethz.ch; erich.meier@geo.uzh.ch

*Author to whom correspondence should be addressed.

ABSTRACT

Spaceborne synthetic aperture radar (SAR) measurements of the Earth's surface depend on electromagnetic waves that are subject to atmospheric path delays, in turn affecting geolocation accuracy. The atmosphere influences radar signal propagation by modifying its velocity and direction, effects which can be modeled. We use TerraSAR-X (TSX) data to investigate improvements in the knowledge of the scene geometry. To precisely estimate atmospheric path delays, we analyse the signal return of four corner reflectors with accurately surveyed positions (based on differential GPS), placed at different altitudes yet with nearly identical slant ranges to the sensor. The comparison of multiple measurements with path delay models under these geometric conditions also makes it possible to evaluate the corrections for the atmospheric path delay made by the TerraSAR processor and to propose possible improvements.

INTRODUCTION

The correction of atmospheric path delays in high-resolution spaceborne synthetic aperture radar systems has become increasingly important with continuing improvements to the resolution of SAR systems surveying the Earth. Atmospheric path delays must be taken into account in order to achieve geolocation accuracies better than 1 meter. These effects are mainly due to ionospheric and tropospheric influences. Path delays through the ionosphere are frequency-dependent, proportional to the inverse square of the carrier [1, 2]. At frequencies higher than L-band under average solar conditions, the major contribution of the atmospheric path delay comes from the troposphere [2, 3]. The tropospheric delay is usually divided into hydrostatic, wet and liquid components [4]. The hydrostatic delay is mainly related to the dependency of the refractive index on the air pressure (i.e. target altitude) and the wet delay on the water vapour pressure. The liquid delay is due to clouds and water droplets. While the wet component can be highly variable, the hydrostatic delay normally only changes marginally because of the lack of significant pressure variations within the extent of a typical SAR scene [4]. Interferometric radar meteorology produces high resolution maps of integrated water vapour for investigations in atmospheric dynamics and forecasting [4]. Using that knowledge, global and local atmospheric effects (e.g. vortex streets, heterogeneities, turbulences) can be detected or even removed using interferometric and multi-temporal data [5–7], or by inclusion of global water vapour maps from the ENVISAT Medium Resolution Imaging Spectrometer (MERIS) sensor [8]. In addition to interferometric applications, there is a growing interest in the correction of atmospheric influences within a single SAR image. Especially for SAR geolocation measurements, these atmospheric contributions introduce 'geolocation noise' that without correction causes shifts in geocoded products.

In this paper, the tropospheric path delay was assumed to depend only on the target's altitude and the local incidence angle of the radar wave. As the variability of the wet path delay is within ¼0.3 m [4], the wet delay in the model is based on average atmospheric conditions, maintaining the heightand incidence angle dependencies. Thus, the contribution of the wet component to the geolocation error should usually be significant below < 0.15 m. For comparison and as a reference model, a ray-tracing approach using current weather data is introduced. A set of TSX data and GPS measurements are used to verify the results from the model, as well as for comparison with the operational TSX processor's own atmospheric correction factors. The ionospheric contributions are estimated using TEC estimates from the GNSS network, and are compared to the DLR processor estimates provided in the TSX products. Since the TSX operational processor corrects the whole scene in question for the influence of the atmosphere using average TEC values, the mean scene height and the nominal mid-range incidenceangle [9], atmosphere-induced geolocation errors of ¼1 m are possible in mountainous regions. Together with DGPS measurements of four on-site corner reflectors and the TSX data, the results from the models and the measurements were cross-validated. A set of six TSX scenes were used to compare the operational 'average' atmospheric correction to a model utilizing meteorological data, as well as to a simple altitude-dependent model. While the meteorological model may not be suitable for operational use, the altitude-dependent model is straightforward and easy to implement. A comparison between these approaches and the DGPS measurements indicates a path toward improvement, especially in mountainous areas.

METHODOLOGY

In the following, six TerraSAR-X Stripmap scenes (30 km x 20 km) containing four identical corner reflectors at altitudes of ¼570 m (Meiringen/Interlaken) and ¼3580 m (Jungfraujoch) were examined. Figure 1 illustrates the geometry and location of the scenes. In order to obtain nearly identical ranges for reflectors at different off-nadir angles, the reflectors closer to nadir are located ¼3000 m below the reflectors farther from nadir. Locations fulfilling these requirements were found in Switzerland for the descending case with a pair covering the Jungfraujoch and Meiringen regions, and for the ascending case with a Jungfraujoch and Interlaken pair. The arrangement serves two purposes:

(1) The same nominal antenna gain pattern correction is normally applied to two equal-range reflectors.

Therefore, differences in their reflected intensities indicate topography-induced antenna gain pattern correction errors (not investigated within this paper).

(2) The nominal correction scheme for the atmospheric path delay can be tested by comparing predicted and measured ranges. The range differences between the high- and low-altitude reflectors help quantify *relative* differences in the path delay.

Another interesting side effect is that the average scene height in both configurations is close to the midpoint between the two reflector altitudes. Additional meteorological data (temperature, water vapour pressure, air pressure) from weather stations near Meiringen, Interlaken and Jungfraujoch provided further reference information for accurate modeling of the refractive index and atmospheric path delays.

Though they play only a minor role in this case, ionospheric path delays observed during the data takes and at the corresponding locations were estimated using the total electron content (TEC) along the ray path. TEC measurements were obtained from global vertical TEC maps with bi-hourly temporal resolution. The TEC maps can be downloaded in the IONosphere map EXchange format (IONEX) from the Center for Orbit Determination in Europe (CODE) [10].

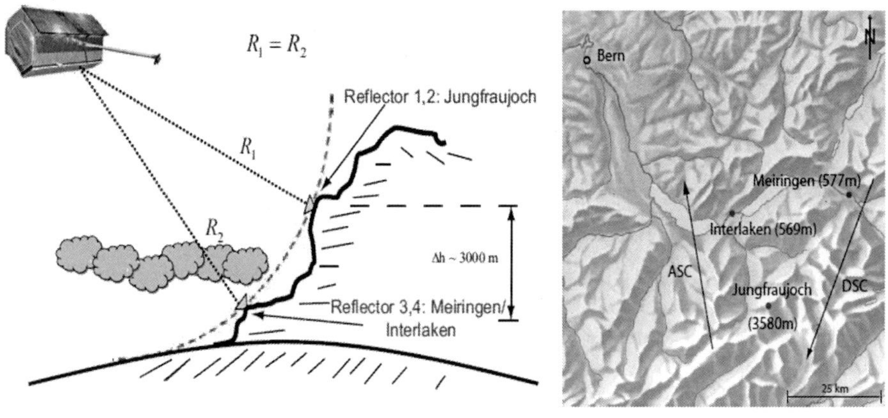

Figure 1. Observation geometry.

MODELS AND MEASUREMENTS

The following sections provide a brief description of two models used:

(a) Raytracer, and (b) heightdependent.

In addition, the measurements made for the estimation of the atmospheric path delays are described. While the raytracer uses weather data for an estimation of the path delays with mm accuracy, the altitude-dependent approach should provide a simplified model to correct path delays with cm accuracy.

Raytracer

The tropospheric delay is estimated on the basis of data provided by a numerical weather model [11]. Using this information, the raytracing algorithm integrates through the refractivity field along the path between the satellite and the point on the surface of the Earth. The non-hydrostatic local area model COSMO-2 is used as a numerical weather model. It is operated by the Swiss Federal Office of Meteorology and Climatology and covers central Europe. It has a resolution of about 2 km and consists of 60 layers. The bottom layer follows the terrain, while the top boundary ends at 23589 m above the reference ellipsoid (WGS84). The model is used for the determination of the refractivity.

The raytracer assumes that the path followed by the ray is equivalent to the shortest geometrical path between the satellite and the point of interest. It is therefore only determined by the satellite position and the point of interest, but not by the refractivity field. This permits a simple computation of the ray paths. Since the refractivity field and its variability decreases with altitude, the length of the integration steps can be enlarged at upper levels without significantly reducing the accuracy, saving computation time. The integration method used is Newton-Cotes quadrature. However, this method has a fixed integration step length. To overcome this constraint, the atmosphere is subdivided into layers, each with a characteristic integration step size. By fixing the number of sampling points and increasing the thickness of a layer, the section of the ray path within the layer is lengthened. Consequently, the step size is increased, slightly reducing the accuracy. The thicknesses of the layers are chosen to cause zenith path delays nearly equivalent to those from a standard refractivity atmosphere (see Equation 2 and [12]).

The boundaries of the layers are computed by following the recursive formula

$$h_{i+1} = -h_s \log(\exp(-\frac{h_i}{h_s}) - \frac{dt_{\text{tot}}}{n}) \tag{1}$$

$$dt_{\text{tot}} = \int_{h_0}^{h_n} \exp(-\frac{h}{h_s})dh \tag{2}$$

where *hi* is the lower boundary height of the *i*-th layer (*i* = 0;….. ; *n* ¡ 1) and *hs* the scale height. The parameter *h*0 is set to the height of the point of interest and *hn* to the height up to which will be integrated. The integration is carried out for each layer, and then the delays for all layers are added up to obtain the total result.

The refractivity N is usually not a prognostic variable in numerical weather models. However, it can be calculated from the partial dry air pressure *pd* (in hPa), temperature *T* (in K) and partial water vapour pressure *pw* (in hPa) using a formula published by R¨ueger [13]

$$N = 77.6890\frac{p_d}{T} + 71.2952\frac{p_w}{T} + 375463\frac{p_w}{T^2}. \quad (3)$$

Table 1. Setup parameters for the integration algorithm.

Parameter	Value
Scale height h_s	7353.0 m
Integration hcight h_n	60000 m
Number of layers n	12
Number of sampling points per layer	12

The prognostic variables used in the integration are interpolated at the sampling points. A multi-linear interpolation method is applied in 4 dimensions (space and time). If a sampling point is located outside the domain of the weather model, the meteorological quantities are extrapolated from the values at the boundaries. Points situated above or below the area of interest are exponentially extrapolated, whereas points located adjacent to the area of interest are set to the value of the nearest boundary point. The values of the interpolation parameters are listed in Table 1. The free parameter *hs* in the exponential function used for estimating the layer thickness is proposed in [12]. The upper bound of the accuracy of the integration algorithm is set to 1 mm. Several tests were carried out to find adequate values for the remaining parameters fulfilling the accuracy constraints. More details on the accuracy of raytracers can be found in [14, 15].

Height-dependent Model

Since the tropospheric delay is most sensitive to altitude, a quick, straightforward and purely heightdependent approach was derived from standard models including mean estimates of the surface airpressure *P0*, temperature and water vapour. As mentioned in the introduction, the tropospheric delay is usually divided into hydrostatic-$(\Psi_{\mathtt{hyd}})$wet $(\Psi_{\mathtt{wet}})$ and liquid $(\Psi_{\mathtt{liq}})$ components and can be written as [16]:

$$\Psi_{\mathtt{tropo}} = \Psi_{\mathtt{hyd}} + \Psi_{\mathtt{wet}} + \Psi_{\mathtt{liq}} \tag{4}$$

The hydrostatic component refers to a standard atmosphere (in hydrostatic equilibrium). The wet component accounts for the water vapour while the liquid component takes into account the liquid water content (clouds, droplets) along the signal path. Due to its small contribution (on the order of a mm) ψ_{liq} is usually neglected for SAR path delay estimates [4]. The hydrostatic component ψ_{hyd} in the nadir direction can be derived from [17, 18]:

$$\Psi_{\mathtt{hyd}} = 10^{-6} k_1 \cdot \frac{R_d}{g_m} P_0. \tag{5}$$

Whereg$_m$istheaccelerationduetolocalgravity, k_1=77.6 $[\frac{K}{\mathtt{mbar}}]$ is a refractive constant, and R_d=287 $[\frac{J}{K \cdot \mathtt{kg}}]$ is the ideal gas constant. The wet delay contribution is estimated using [19]:

$$\Psi_{\mathtt{wet}} = 10^{-6} \cdot \left(\frac{(k_2' T_m + k_3) R_d e_0}{T_0(g_m(\lambda + 1) - \beta R_d)} \right) \cdot \kappa_{wet} \tag{6}$$

With

$$\kappa_{\mathtt{wet}} = \left(1 - \frac{\beta h}{T_0}\right)^{\frac{(\lambda+1)gm}{R_d\beta}-1} \quad (7)$$

Table 2. Parameters used to model the tropospheric path delay.

$P_{0,\mathtt{mid}}$	$T_{0,\mathtt{mid}}$	$e_{0,\mathtt{mid}}$	$\beta_{\mathtt{mid}}$	$\lambda_{\mathtt{mid}}$
1013.25 [hPa]	288.15 [K]	11.691 [hPa]	$6.5 \cdot 10^{-3}$ [K/m]	3 [-]

Where $k_2' = 23.3 \left[\frac{K}{\mathtt{mbar}}\right]$, $k_3 = 3.75 \cdot 10^5 \left[\frac{K^2}{\mathtt{mbar}}\right]$ are refractive constants $\beta = 6.5 \left[\frac{K}{\mathrm{km}}\right]$ is the temperature lapse rate, T_0 [K] the temperature e_0 [hPa] the water vapour pressure above sea level, *Tm* [K] the mean temperature of water vapour, *h* the target height and ¸ [unitless] the average water vapour decrease. For the generation of a single altitude-dependent estimate of the tropospheric path delay for SAR applications, the average atmospheric parameters shown in Table 2 are used to determine the coefficients of a polynomial that fits the path delay data for heights h ranging from 0 to 9000 m in a least-squares sense. The resulting approach

$$\Psi_{\mathtt{tropo,zenith}} = \frac{h^2}{8.55 \cdot 10^7} - \frac{h}{3411} + 2.41 \quad [\mathrm{m}] \quad (8)$$

estimates the tropospheric delay in the zenith direction. The delay in the look direction of the antenna can be approximated using the nominal incidence angle α_{inc} [rad] according to

$$\Psi_{\mathrm{tropo}} = \frac{\Psi_{\mathrm{tropo,\,zenith}}}{\cos \alpha_{\mathrm{inc}}} \quad [\mathrm{m}], \tag{9}$$

where the incidence angle is calculated using the local height above the ellipsoid. This approximation of tropospheric path delay is compared later with an average delay correction and a path delay estimation based on the raytracer using meteorological data as described above. Figure 2 shows a block diagram of the three methods used for the estimation of tropospheric path delays. Figure 2a) illustrates the altitude-dependent method, Figure 2b) the raytracer and Figure 2c) the estimation and comparison of path delays from SAR data and GPS measurements described in the following section (JJ: Jungfraujoch, MI: Meiringen/Interlaken).

SAR and GPS Measurements

Estimation of atmospheric path delay directly from the SAR data was performed using precise DGPS measurements of four observed targets (corner reflectors) in the images. The range distance between the DGPS coordinates and the sensor is considered to be the reference for all estimates. The atmospheric corrections proposed in the TSX annotations, based on an average reference height (AVG), are applied to the image data (correction of fast time parameters). Their differences compared to the GPS measurements provide relative deviations from the reference range distance. The relative differences in path delay estimates between the corner reflectors located at different altitudes are calculated in a final step. These calculations point out varying propagation properties and indicate possibilities for refinement (Figure 2c). Ionospheric contributions are also considered, although they play a minor role at X-band frequencies, especially at the current solar minimum. Average estimates for their contribution (path delay at 5 TECU) are provided in the TSX product annotations [20]. Measured values were estimated using total electron content (TEC) maps describing spatial variations in TEC values across the Earth. The one-way ionospheric path delay may be estimated from [21]:

$$\Psi_{\text{iono}} = K \cdot \frac{\text{TEC}}{f_c^2} \cdot \frac{1}{\cos \alpha_{\text{inc}}} \tag{10}$$

where *fc* is the center frequency of the radar wave, *c* the speed of light and $K = 40.3 \frac{\text{m}^3}{\text{s}^2}$ is a refractive constant. The factor $\frac{1}{\cos \alpha_{\text{inc}}}$ converts the path delay from nadir to the path at a particular incidenceangle.

Measurements from GNSS networks provide multiple maps per day, and may be downloaded in a standardized format from the internet [10]. Given the TEC values together with the corresponding satellite and target positions, the expected path delay is calculated.

Path Delay Results

As a first test, the absolute image localization error for all four corner reflectors in each of the six TSX products was measured. Accurately surveyed DGPS measurements of the corner reflectors were used to predict their range and azimuth positions in each image, and these predictions were compared to their measured locations. Figure 3 shows an example for the absolute location error estimate for (a) Jungfraujoch (JJ), (b) Interlaken (INT) and (c) Meiringen (MEI). At successive 'zoom' levels, the blue crosses indicate the prediction based on the GPS measurements, which represent the runtime measurements under ideal vacuum conditions. These image position predictions were made by solving the Doppler and range equations [22] using the surveyed target coordinates, together with the precise orbit state vectors and the image timing annotations [20]. The precise corner reflector (phase center) position in the image was determined by searching for local maxima in the neighborhood of the strong targets, using complex-FFT oversampling (factor of 50) to obtain sub-sample accuracy [22]. Results of the following analysis are based on the location error of the two corner reflectors at each test site, both at the same altitude and equidistant to the sensor. Figure 4 shows a scatter plot of all estimated location errors. The blue circles indicate descending-, the red circles ascending products. While for Interlaken and Meiringen (b) the range errors are on the order of a cm, for the Jungfraujoch site (a) the range errors increase to a mean of approximately 0.58 m. Since the TSX tropospheric correction is based on an *average* scene height roughly halfway between the test site altitudes (Jungfraujoch and Meiringen/Interlaken), the expected average location errors in range for the sites would be expected be approximately of the same magnitude with opposite signs.

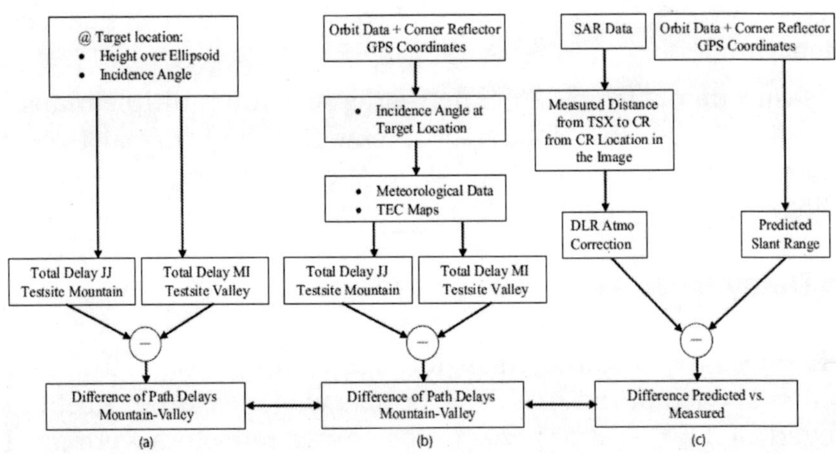

Figure 2. Scheme of methods used for comparison of atmospheric path delays; Testsite Jungfraujoch (JJ) on mountain site, Meiringen and Interlaken (MI) in the valley.

In order to estimate the height-dependent path delays for each test site, the range errors were subtracted from the average range delay (see Table 3). As expected, atmospheric delays at the higher altitude of the Jungfraujoch site (y_{JJ},AVG) are smaller than at the

Meiringen/Interlaken sites (y_{MI},AVG). The estimated relative difference in one-way path delay between both sites has a mean of 0.779 m. Differences in the descending case are usually higher in comparison to the ascending cases, as the signal path through the troposphere was longer, due to the more oblique incidence angle. Results from the height dependent model (HM) and the raytracer (RT) use the altitudes and incidence angles estimated for each corner reflector position. The path delays from the TSX annotations (ϕyAVG) are used together with the differences ΔRg.JJ and ΔRg.MI to map to the delays according to the altitude of the testsite. Figure 5(a) and (b) show the total tropospheric path delays (hydro + wet component) estimated from the models, including the results from the measurements as well as the ionospheric delays.

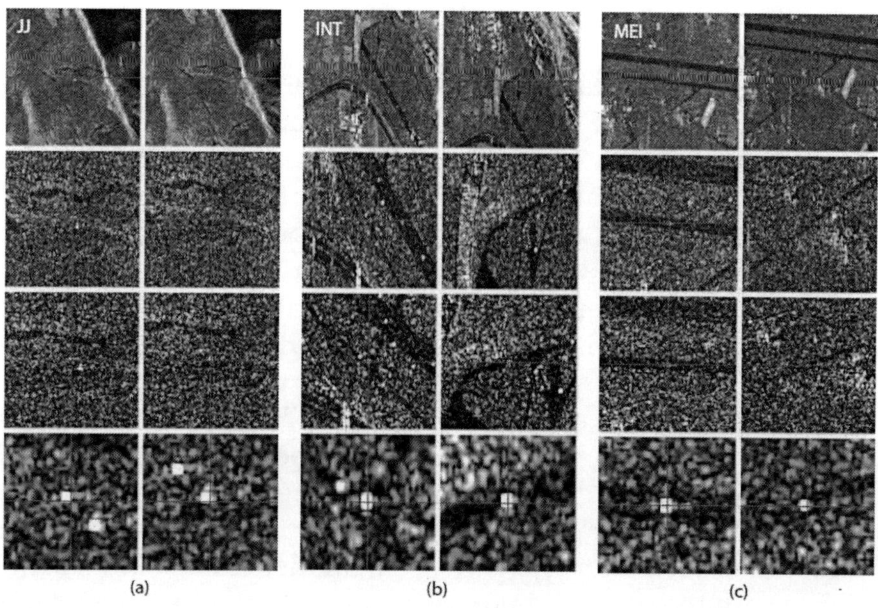

Figure 3. Example of zoomed in and interpolated corner reflectors in TSX imagery. The blue cross indicates the GPS derived position prediction and the strong white target the actual measured position of the CR in the image.

Table 3. Predicted path delays from measurements and GPS. The ϕRef.Height, αinc and ϕyAVG refer to the average scene height, the mid incidence angle and average tropospheric path delay respectively, as annotated in the TSX products (JJ: Jungfraujoch, MI: Meiringen/Interlaken, dates are all in the year 2008).

Predicted Path Delay from Measurement and GPS									
Date	A/D	ØRef.Height	α_{inc}	$\varnothing\Psi_{AVG}$	ΔRg_{JJ}	ΔRg_{MI}	$\Psi_{JJ,AVG}$	$\Psi_{MI,AVG}$	Ψ_{JJ}-Ψ_{MI}
28.04.	D	2163 m	31.2°	1.874 m	0.846 m	0.036 m	1.03 m	1.84 m	0.809 m
09.05.	D	2166 m	31.2°	1.873 m	0.818 m	-0.064 m	1.05 m	1.94 m	0.882 m
12.05.	A	1865 m	24.0°	1.845 m	0.755 m	0.055 m	1.09 m	1.79 m	0.700 m
23.05.	A	1827 m	24.0°	1.857 m	0.782 m	0.068 m	1.07 m	1.79 m	0.714 m
11.06.	D	2164 m	31.2°	1.874 m	0.646 m	-0.091 m	1.23 m	1.96 m	0.737 m
25.06.	A	1827 m	24.0°	1.857 m	0.418 m	-0.414 m	1.44 m	2.27 m	0.832 m
Mean Values				1.863 m	0.581 m	-0.068 m	1.16 m	1.92 m	0.779 m

In Figure 5(c) the black and yellow lines show the distribution of air pressure and water vapour during the data takes in comparison to the wet path delays estimated from the raytracer. Both pressure parameters are normalized to the assumptions of the standard atmosphere made in Table 2. Figure 5(d) shows the differences in atmospheric path between the mountain (Jungfraujoch) and the valley (Meiringen/Interlaken) testsites and therefore measures the dynamic of the path delay models at these altitudes. The results from the height-dependent model are very similar to the raytracer results. An exceptional scene was the data from June 25th (heavy rain) where high water vapour pressure was measured, which significantly increased the wet path delay contribution. The path delays estimated from the image as compared to the GPS measurements are plotted in red, and are less consistent with the model results, but strongly correlate with the water vapour measurements (Figures 5(a) to (c)).

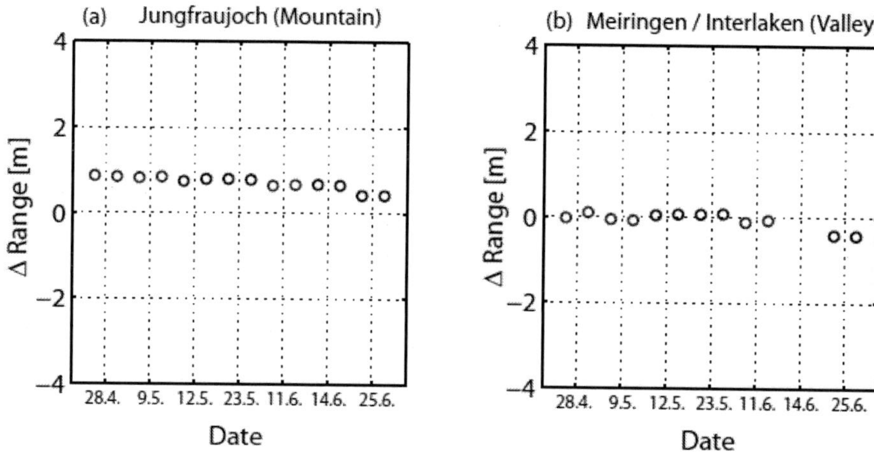

Figure 4. Absolute location error estimations for each testsite. Red circles mark descending and blue circles ascending geometry. Note: An additional data take is plotted in Figure 4(a). As it was not possible to deploy the CRs in Interlaken at that time, this data take was omitted in the analysis that followed.

This is unsurprising, as the variations arise from the differences between these constant delays and the GPS-measured vacuum propagation (ΔRg.MI, ΔRg.JJ). The TSX average correction does not significantly vary across the ascending/descending geometries and therefore causes only a constant shift. From Figure 5(a) and (b) it can be seen that the shifts between the model and the measurements are nearly constant. This suggests that the tropospheric path delays from the TSX annotations consistently *underestimate* the true delays. On the other hand, this shift implies that if the delays from the models were applied to the TSX image, we would expect an average range shift ΔRg_{MI} and ΔRg_{JJ} of ¼0.7 m across all test sites. It follows that the measurements in Interlaken only performed with cm accuracy since the underestimated path delay having been compensated by the overall constant ¼0.7 m shift.

Path delays caused by the ionosphere were modelled to have been in the cm range. The average one-way ionospheric delays from measurements are comparable to, although slightly higher than, the average values provided in the TSX annotations

$$(\varnothing\Psi_{\text{Meas, iono}}=4.7\ \text{cm},\ \varnothing\Psi_{\text{AVG, iono}}=2.5\ \text{cm})$$

The relative differences of the path delays between the test sites in Figure 5(d) can be considered an indication of the model's ability to capture altitude-dependent variations. With mean values of $\varnothing\Delta\Psi_{\text{HM}}=0.822\ \text{m}$, $\varnothing\Delta\Psi_{\text{RT}}=0.786\ \text{m}$ and $\varnothing\Delta\Psi_{\text{AVG}}=0.779\ \text{m}$ all models agreed closely.

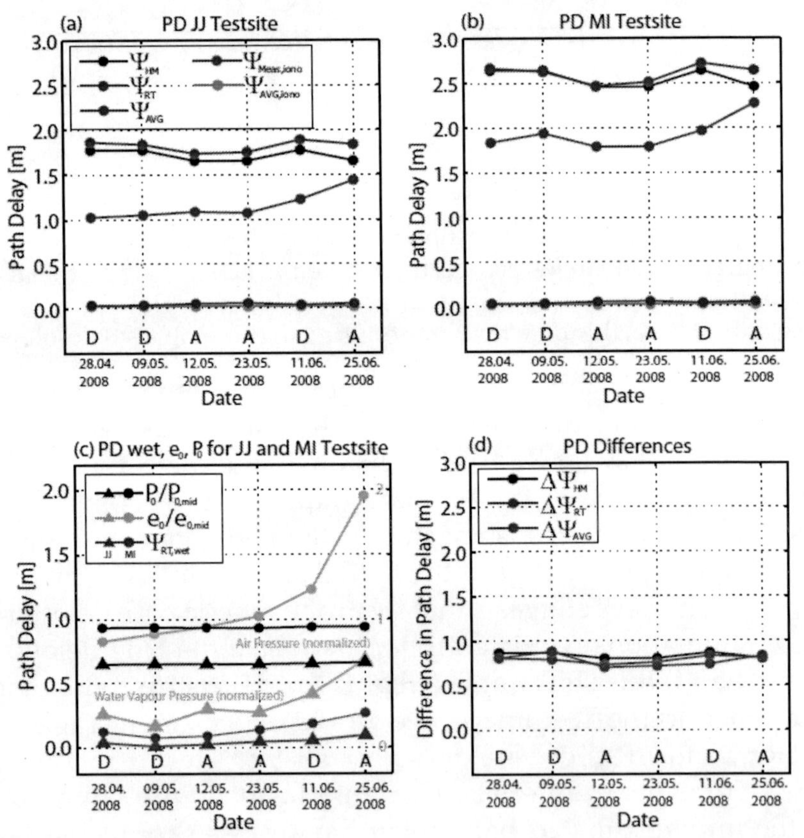

Figure 5. Modelled and measured atmospheric path delays. Path delays in a) for JJ testsite and in b) for the MI testsites. In c) wet path delays compared to measured air- and water vapour pressure (normalized) and in d) path delay differences between the results of the JJ and the MI testsites.

DISCUSSION AND CONCLUSION

With TerraSAR-X, a civilian spaceborne satellite is for the first time able to observe the Earth with a radar resolution on the order of 1 m. At such resolutions, the influence of the atmosphere on the geolocation accuracy plays a significant role and must be taken into account. Various methods for the correction of these delays can be applied, depending on the scientific application. In this work, we investigated the range delays predicted by an altitude- and incidence-angle dependent model, a raytracer and a model which corrects the path delay as a scene average. Model results were compared for six TerraSAR-X datasets containing four corner reflectors, two at altitudes 3000 m above the others.

We found that the scene-average method significantly *underestimates* the tropospheric delays compared to the raytracer model and the height-dependent polynomial. While the range location errors for the low-altitude reflectors were normally on the order of a cm, the range errors at the high altitude station were usually over half a meter. The absolute location error for a scene-average path delay estimate should result in comparable range errors at equal height offsets above and below the average scene height. Our test sites at Jungfraujoch (3580 m) and Meiringen/Interlaken (¼570 m) with average scene heights of 2160 m/1860 m nearly fullfilled that condition. When the annotated delays from the TSX products were replaced by the results from the raytracer for the individual locations, a nearly constant range shift of ¼0.7 m was estimated. On the one hand, this indicates that the corrections from the raytracer are reasonable, yet on the other hand, suggests an inherent systematic shift of ¼0.7 m which is still within the specifications of the TSX accuracy requirements.

Since the dynamics of all methods are similar, differences between the models are likely due to different atmospheric starting positions. The results for the last two acquisitions are more similar for the various models. The range deviations from the image measurements and the path delays from the raytracer show that this is probably due to the increased amount of atmospheric water vapour during these data takes. As a result, the path delay in the SAR image increases. The standard correction from the TSX annotations does not take these effects into account. Therefore, the influence of the higher water vapour pressure could be directly observed in the changes in the range location errors. In other words, the larger path delay in the image compensates for the underestimated average path delay which leads to the observed 'increased' accuracy. Path delay estimates from the raytracer include the higher water vapour pressure, but probably to a lesser extent, as the heavy rainfall was observed to be very localized. This might also be seen from its path delay estimates. Loss of accuracy caused by poor modelling of wet path delay estimation in such storm events are not expected to exceed ¼10 to 15 cm. Separation of the ionospheric delays from the total atmospheric delays also indicates that the average value from the TSX annotations tends to underestimate the delay estimated from the measured TEC maps by a mean of 2.2 cm. Although solar activity and therefore the ionospheric delay is presently near its minimum, future path delays at X-band in mid-latitude regions are not expected to regularly exceed Unlike the scene-average model, the results from the height-dependent polynomial are very close to the raytracer estimates. The height-dependent polynomial would therefore seem to be a straightforward alternative for operational use. Worse results would be expected at high latitudes, as the correction is based on a standard atmosphere most representative of mid-latitude regions. This could be improved relatively simply with extra terms 'capturing' latitude-induced dynamics. As the raytracer produces the most accurate results, a correction of the data with this approach would be desirable; however, operational inclusion of the current weather data is currently not feasible.

ACKNOWLEDGMENTS

We would like to thank the German Aerospace Center (DLR) in Oberpfaffenhofen for their scientific support and for providing us with the TerraSAR-X datasets granted within the framework of CAL0163.

Furthermore, we are indebted to the International Foundation High Altitude Research Stations Jungfraujoch and Gornergrat (HFSJG), 3012 Bern, Switzerland, who made it possible for us to carry out our experiments at Jungfraujoch. We also would like to thank the respective airport authorities for their permission to deploy the corner reflectors on the grounds of the Meiringen and Interlaken airports and Dr. D. Leuenberger working at the Swiss Federal Office of Meteorology and Climatology MeteoSwiss for providing NWP model data.

REFERENCES AND NOTES

1. Davies, K. *Ionospheric Radio;* Peter Peregrinus Ltd., IEE Electromagnetic Waves: London, UK, 1990; Vol. 31. 1.
2. Xu, Z.-W.; Wu, J.; Wu, Z.-S. A survey of ionospheric effects on space-based radar. *Waves in Random Media* 2004, *14,* 189–272. 1.
3. Jehle, M.; Frey, O.; Small, D.; Meier, E.; N¨uesch, D. Estimation and correction of atmospheric path delay in radar signal propagation. Technical Note TX-PGS-TN-3016, Remote Sensing Laboratories, University of Zurich, 2004. 1.
4. Hanssen, R. F. *Radar Interferometry;* Kluwer Academic Publishers: Dordrecht, The Netherlands, 2001; Vol. 2. 1., 3.2.
5. Chaabane, F.; Avallone, A.; Tupin, F.; Briole, P.; Maitre, H. A multitemporal method for correction of tropospheric effects in differential SAR interferometry: Application to the gulf of Corinth earthquake. *IEEE Trans. Geosci. Remote Sens.* 2007, *45,* 1605–1615. 1.
6. Tarayre, H.; Massonnet, D. Effects of a refractive atmosphere on interferometric processing. In *Proc. Geosci. Remote Sens. Symp., 1994. IGARSS '94. Surface and Atmospheric Remote Sensing: Technologies, Data Analysis and Interpretation., International,* Aug 8-12, 1994; Institute of Electrical and Electronics Engineers: Los Alamitos, CA, 1994; Vol. 2, pp. 717–719.
7. Kimura, H.; Kinoshita, H. Effect of tropospheric range delay corrections on differential SAR interferograms. In *Proc. Geosci. Remote Sens. Symp., 2001. IGARSS '01. IEEE 2001 International,* Sydney, Australia, 2001; Institute of Electrical and Electronics Engineers: Los Alamitos, CA, 2001; Vol. 5, pp. 2049–2051. 1.
8. Zeng, Q.; Li, Y.; Li, X. Correction of tropospheric water vapour effect on ASAR interferogram using synchronous MERIS data. In *Proc. Geosci. Remote Sens. Symp., 2007. IGARSS '07. IEEE 2007 International,* Barcelona, Spain, Jul 23-28, 2007; Institute of Electrical and Electronics Engineers: Los Alamitos, CA, 2007; pp. 2086–2089. 1.
9. Fritz, T. Comments on TerraSAR-X Geolocation / Coordinate Annotation (Personal communication and handouts), 2008. 1.
10. Astronomical Institute of the University of Bern. Centre for Orbit Determintation (CODE). http://www-.aiub.unibe.ch/download/CODE/, 2008. 2., 3.3.
11. Doms, G.; Sch¨attler, U. *A description of the nonhydrostatic regional model LM. part I: Dynamics and numerics;* Consortium for Small-Scale Modelling COSMO: Offenbach, Germany, 2002; Report No. LM-F90 2.18, 140 p. 3.1.

12. Recommendations and Reports of the CCIR XVIth Plenary Assembly. Radiometeorological data. volume 5, (Propagation in non-ionized media). Dubrovnik, 1986. 3.1., 3.1.

13. R¨ueger, J. M. Refractive index formulae for radio waves. In *Integration of Techniques and Corrections to Achieve Accurate Engineering*, Proc. XXII FIG Int. Congress ACSM/SPRS Ann. Conf., Washington DC, USA, Apr 19-26, 2002. 3.1.

14. Pany, T.; Pesec, P.; Stangl, G. Atmospheric GPS slant path delays and ray tracing through numerical weather models, a comparison. *Phys. Chem. Earth, Part A: Solid Earth and Geodesy* 2001, *26*, 183–188. 3.1.

15. Yang, X.; Sass, B. H.; Elgered, G.; Johansson, J. M.; Emardson, T. R. A comparison of precipitable water vapor estimates by an NWP simulation and GPS observations. *J. Appl. Meteorol.* 1999, *38*, 941–956. 3.1.

16. Jehle, M.; Small, D.; Meier, E.; N¨uesch, D. Improved knowledge of SAR geometry through atmospheric modelling. In *Proc. EUSAR*, 2004. 3.2.

17. Saastamoinen, J. Contributions to the theory of atmospheric refraction. *Bull. Geodes.* 1973, 105-107. 3.2.

18. Davis, J.; Herring, T.; Shapiro, I.; A.E.E.Rogers.; Elgered, G. Geodesy by radio interferometry: Effects of atmospheric modelling errors on estimates of baseline length. *Radio Sci.* 1985, *20*, 1593–1607. 3.2.

19. Collins, P.; Langley, R. Limiting factors in tropospheric propagation delay error modelling for GPS airborne navigation. In *Proc. Inst. Navig. 52nd Ann. Meet.* 1996. 3.2.

20. Fritz, T. TerraSAR-X ground segment. Level 1b Product Format Specification TX-GS-DD-3307, DLR, 2007. 3.3., 4.21. Wild, U. Ionosphere and geodedetic satellite systems: Permanent GPS tracking data for modelling and monitoring. *Geod¨atisch-geophysikalische Arbeiten in der Schweiz* 1994, *48*, 155 p. 3.3.

21. Small, D.; Rosich, B.; Schubert, A.; Meier, E.; N¨uesch, D. Geometric validation of low and high resolution ASAR imagery. In *Proc. 2004 Envisat ERS Symp.*, Salzburg, Austria, 2005; No. ESA SP-572, p. 8. 4

Chapter 7

ANALYSIS OF RFI IDENTIFICATION AND MITIGATION IN CAROLS RADIOMETER DATA USING A HARDWARE SPECTRUM ANALYSER

Pascal Fanise [1], Mickaël Pardé [1], Mehrez Zribi [2,*]
Monique Dechambre [1] and Christophe Caudoux [1]

[1] LATMOS, Route de Troux, 78280 Guyancourt, France

[2] CESBIO (UPS/CNRS/IRD/CNES), 18 Avenue Edouard Belin, bpi 2801, 31401 Toulouse Cedex 9, France

ABSTRACT

A method to identify and mitigate radio frequency interference (RFI) in microwave radiometry based on the use of a spectrum analyzer has been developed. This method has been tested with CAROLS L-band airborne radiometer data that are strongly corrupted by RFI. RFI is a major limiting factor in passive microwave remote sensing interpretation. Although the 1.400–1.427 GHz bandwidth is protected, RFI sources close to these frequencies are still capable of corrupting radiometric measurements. In order to reduce the detrimental effects of RFI on brightness temperature measurements, a new spectrum analyzer has been added to the CAROLS radiometer system. A post processing algorithm is proposed, based on selective filters within the useful bandwidth divided into sub-bands. Two discriminant analyses based on the computation of kurtosis and Euclidian distances have been compared evaluated and validated in order to accurately separate the RF interference from natural signals.

INTRODUCTION

In recent years, several studies have revealed the highly detrimental influence of radio frequency interference (RFI) on microwave observations of the Earth, especially in the case of spatialradiometry [1]. This kind of perturbation corrupts the recorded signals, thus deteriorating the data quality to variable degrees which, in some cases, render it unusable. The presence of strong RFI is particularly noticeable over continental surfaces, but has also been observed by the recently launched SMOS satellite over the ocean [2].

At low frequencies (L-band), the soil moisture (SM) fraction has a negative-signed dependence on the brightness temperature (Tb) of 1 K/%, with its exact value depending on soil type, vegetation cover, surface roughness and other factors. The SMOS radiometer has been designed to measure the surface soil moisture, with a specified radiometric error of less than 2 K, depending on the nature of the target and its position in the instrument field of view. The SM retrieval error is predicted to be lower than 4%, as required by the hydrology community [2]. Small levels of RFI, producing only a few Kelvin in terms of equivalent radiometric signal, would thus cause the soil to appear dryer, which in turn would be interpreted as lower recent rainfall, or higher recent evaporation. The large spatial scale of satellite measurements could be a second factor contributing to an increase in the presence of RFI.

The future SMAP satellite [3] will benefit from current RFI studies: this mission is designed to analyse the full protected bandwidth (which cannot be done by SMOS), in order to mitigate RFI effects on brightness temperature (Tb) measurements, by identifying and eliminating the sub-bands which are affected by RFI.

RFI detection and mitigation have already been studied and reported in several papers, most of them are related to radio astronomy [4] or C-band [5] and L-band [6-8] Earth surface remote sensing; among those studies, several are based on the analysis of real airborne data [9-11]. Recently, different methods were developed and evaluated, based on the identification of RFI in radiometric measurements. The analysis of kurtosis thresholds applied to Tb measurements [12], or to the first and second U, V Stokes parameters [9,10], has also been evaluated for contaminated data detection in the time domain.

An alternative approach, in the frequency domain, is based on the identification of RFI in the radiometric spectrum [13-15], followed, if possible, by a spectral blanking, *i.e.*, the elimination of the corresponding contaminated data from the spectrum. It is hoped that this approach, which is an attempt to preserve only the cleaned data, will drastically reduce the amount of radiometric data lost in spatial missions such as SMAP.

Various studies based on the spectral analysis of RFI signals have been made in the past. In [6] the authors demonstrated the ability of an ADD (Agile Digital Detector) to detect and remove RFI from microwave measurements. The ADD performance was experimentally verified under controlled laboratory conditions, and in the field near to commercial air traffic control radar. In [7], the authors tested a double RFI detector based on a kurtosis parameter computation, under equivalent conditions. These tests have been conducted with data acquired from ground-based radiometers or from laboratory measurements. In addition, recent airborne radiometer campaigns were carried out with the aim of RFI surveys in the USA [16].

In France, in the context of validation campaigns for the SMOS mission, the CAROLS L-band radiometer [17] was installed onboard the French ATR-42 research aircraft and operated during the 2007 to 2010 —CAROLS campaigns□. These first campaigns were dedicated to the qualification and certification of the instrument and to the validation of the data quality, and, at the end, to the acquisition of different types of brightness measurements over oceanic and land surfaces (in the south west of France). During all the different flights, mainly over the land, we observed the presence of high level RFI perturbations, continuous as well as pulsed. These perturbations were not really expected because of the supposed protected observation bandwidth. We analyzed the corrupted signals and attempted to mitigate them [10]. In addition to our analysis, we decided to build and add the CAROLS airborne radiometer, and operated this new system during the last flight of the last campaign. In this paper, we present a study that we have conducted in order the analyze the RFI present in the L-band spectrum (1,400–1,427 MHz) using this small but helpful database. The aim of this study was the validation of the new analyzer rather than the validation of the algorithms used to interpret the data, and more particularly the data acquired close to high RFI areas

Although instruments obtaining similar spectrally-resolved data have been deployed in previous more extensive airborne campaigns, the results to be shown, though limited, are of interest because they are measured in geographic areas in that have note been studied previously with a radiometer having spectral resolution.

Our paper is organized as follows: Section 2 presents the radiometer and spectrum analyzer system, Section 3 presents our analysis of the radiometric data, and more particularly the data acquired close to high RFI areas, and Section 4 presents two different RFI mitigation algorithms based on subband filtering. All of these were tested with the CAROLS 2009 database, the results of which are presented in Section 5. Finally, our results are discussed in Section 6.

THE CAROLS RADIOMETER AND THE ADDED SPECTRUM ANALYZER SYSTEM

CAROLS Radiometer

The receiver was designed and built as a copy of the EMIRADII [18] radiometer constructed by the DTU (Danish Technical University) team, and was adapted to the French ATR42 research aircraft. It is a fully polarimetric correlation radiometer with direct sampling (using the subsampling concept described in [18]), enabling the measurements of the four Stokes parameters describing the observed electric fields. The detailed characteristics of CAROLS were presented in [17], and only its main characteristics are summarized here. The radiometer is calibrated using internal loads and a pulsed noise diode, which adds approximately 150 K to its input. This setup ensures the system ability to carry out regular internal calibrations, and allows the radiometer to calibrate any phase difference between the two input channels preceding the digital correlator.

Table 1. Specifications of the CAROLS radiometer.

Radiometer	Total power
Frequency	1.400–1.427 GHz (−60 dB bandwidth)
Frequency sampling	139.4 MHz
Stability	0.1 K over 15 mn
Sensitivity	0.1 K for 1 s integration time
Integration	1 ms and 1.8 μs
Calibration	Internal load and noise diode
Antenna	Potter horn (no sidelobe) with 37.6° HPBW
Temperature regulation (analog front end)	45 °C ± 0.1 K

The stability, accuracy and linearity of the radiometer were fully analyzed in the laboratory (see Zribi *et al.*, [17]) using a cryogenic load, and lead to the specifications described in Table 1.

Spectrum Analyzer System

The architecture of the spectrum analyzer (SA) developed for the CAROLS campaigns is presented in Figure 1. The first element is a high resolution analog to digital converter (ADC), followed by a field programmable gate array (FPGA) whose role is to perform digital signal processing.

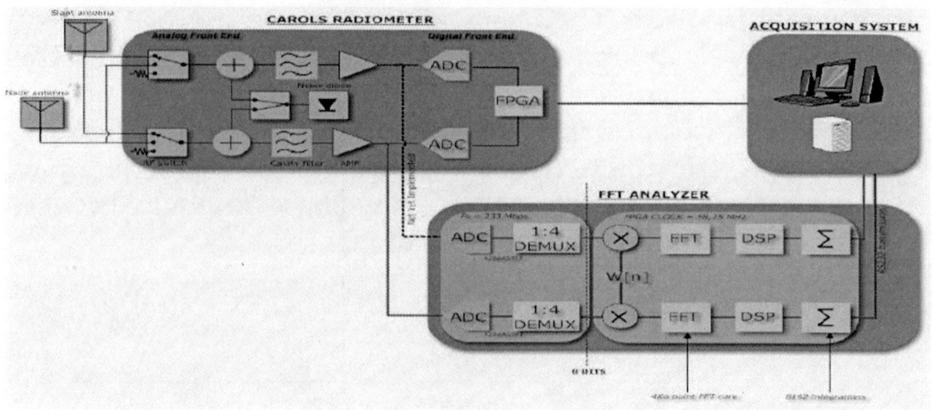

Figure 1. Block diagram of the CAROLS radiometer and (FFT) spectrum analyzer. The red connection was not fully operational during the 2009 CAROLS campaign.

The SA uses the principle of direct RF sampling. The analog signal is bandwidth limited and it can be sampled and digitized immediately after the front end amplification and filtering stages. This technique is used to replace the use of mixer and oscillator local by a very large wideband analog to digital converter (ADC). Then, the band of interest (1.4–1.427 GHz) can be translated down to 2–29 MHz by using a frequency clock to 233 MHz. The converter clock has been carefully chosen with a phase noise below 150 dBc/Hz (@ 10 MHz) to conform to the ADC specification. The de-multiplexed output is allowed to keep only one sample out of four after digitization. The RF signal can be perfectly reconstructed, without loss of information, if the sampling frequency is greater than twice the bandwidth of the signal being sampled, according to Nyquist principle.

Following this first step, the demux output is multiplied by a symmetrical window function (*i.e.*, a Hamming function) to prevent truncation artifacts. The digital signal processing consists in the computation of a Fast Fourier transform (FFT), followed by the power spectrum estimation and accumulation. The sampling rate into the FPGA is set to 58.25 MHz. A 4K FFT process takes place into the FPGA, each FFT is then performed in 70.32 μs. The frequency resolution is 14.22 KHz. We have chosen 1,024 integration, the FPGA was providing a —mean FFT□ every 72 ms.

Connected to the H channel of the CAROLS radiometer slant antenna during the present campaign, the SA allows one to record the modulus of integrated spectra, according to the specifications described in Table 2 below.

Table 2. Spectrum analyzer specifications.

Frequency bandwidth	1.3978–1.4268 GHz
Number of integrations	1,024
Recording time	72.01 ms
FFT	4K points
Windowing	Hamming
Resolution of the FFT	14.16 kHz

PRESENTATION OF THE RADIOMETRIC DATA AND SPECTRA

CAROLS Flight

A special flight of the ATR42 research aircraft, operated by SAFIRE, was carried out on 29 May 2009. Figure 2 illustrates the flight path over southwest France. The aircraft flew at a constant altitude of 3,000 m above the ground between Toulouse and the Atlantic Ocean which are about 235 km apart. A large amount of measurements made along the transect was disturbed by interference in both H and V polarizations [10]. Pulsed RFI sources appeared clearly during most of the flight, as well as two point-like sources producing continuous-wave (CW) RFI, which strongly disturbed the radiometric measurements. The latter sources were situated close to the city of Auch, and to the military ―Centre d'Essai des Landes□ (CEL, *Moorlands Test Centre*) area in the vicinity of the western extremity of the flight path (south of Bordeaux).

Figure 2. CAROLS flight path on 29 May 2009. The portion of the flight indicated by the red arrow was dedicated to the use of a spectrum analyzer, in combination with the CAROLS radiometer.

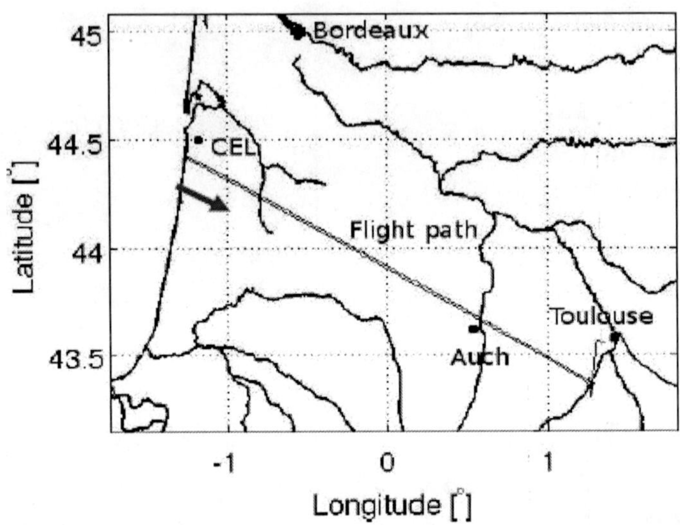

During this flight, radiometric spectra were acquired during both legs of the journey (from Toulouse to the Atlantic and from the Atlantic back to Toulouse), using two sampling frequencies. For the purposes of this study, we chose to analyze the data corresponding to the second part of the flight only, in order to ensure homogeneity of the analysis. Moreover, this leg of the flight was more strongly disturbed by RFI, as the slant antenna was facing towards the south where the sources were situated.

Data from the Spectrum Analyzer

Figure 3 represents the data collected during the spectral analysis flight. To plot this figure, each frequency channel was first calibrated: we used samples acquired regularly looking to an internal load source (~310 K), with and without an internal noise source (adding 140 K), which allow us to have two points for calibration line definition and then Tb estimation. An external cryogenic load was used for the noise source evaluation and for cable loses estimation.

The spectral variations were plotted as a function of time from the ocean to Toulouse (from the left to the right of the figure).

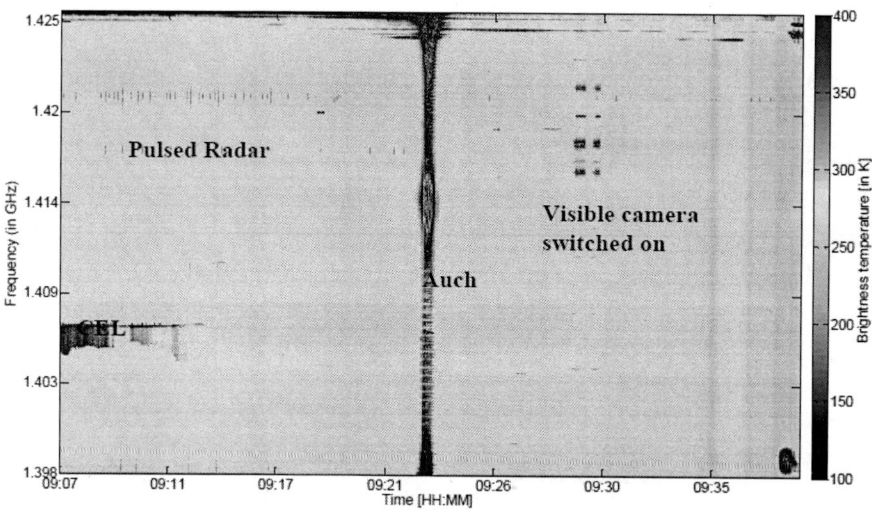

Figure 3. Time-Frequency representation of calibrated Tb values, during the flight from the Atlantic Ocean to Toulouse city (France).

During land survey several peaks corresponding to the presence of three distinct types of RFI can be identified: pulsed RFI sources in the frequency range 1.418–1.419 GHz, emitted by military and civil radars, CW RFI in a narrow band (CEL), and CW RFI occupying the full bandwidth (Auch). Because the latter form of RFI is distributed over the full bandwidth of interest, it was not possible to restore the correct values for Tb. Between 9:07 and 9:23, disturbance is present near 1.426 GHz, probably due to the antennas situated in the vicinity of Auch. These antennas were identified and are known to have a central emission frequencies between 1,427.1 and 1,427.9 MHz, with H or V polarization.

Additional disturbances can be seen between 9:28 and 9:30, corresponding to the switching on and off of a visible camera, thus revealing the strong sensitivity of radiometric measurements to other onboard instruments. Moreover, vertical and horizontal strips are observed on Figure 3 independently to the RFI. The vertical ones, which are not regularly spaced in time, are probably due to intrinsic variations of natural Tb. The horizontal ones are due to the calibration versus frequencies which varies slightly from one frequency to another.

Some other features can be seen on this figure where no obvious RFI is present. Some of these vertical features are due to natural variations; the horizontal one could be due to the independent calibration for each channel.

In Figure 4(a), a pulsed RFI source is represented during the ocean survey—just before the beginning of the transect shown in Figure 3 (we selected these spectrum over sea because it is easier to discriminate small pulsed RFI from natural emission). A peak of 50 K, centered on 1.418 GHz, is clearly visible. This pulse occurred several times during the flight, and its intermittent spectrum can be seen in the left half of Figure 3 at the same frequency.

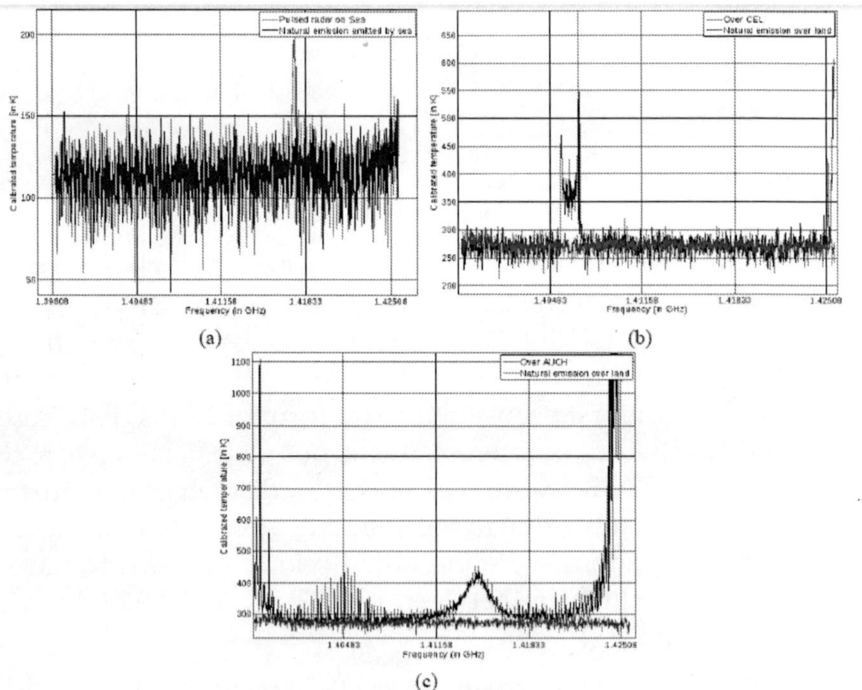

Figure 4. Spectral samples corresponding to **(a)** Pulsed RFI, **(b)** full band CW RFI and **(c)** narrow band CEL RFI. In each figure, the red curve corresponds to the unpolluted (natural) radiometric spectrum.

Figure 4(b,c) illustrates calibrated temperature values as a function of frequency, for the two areas corresponding to Auch and the CEL, respectively. Concerning the perturbations detected in the Auch area, strong RFI can be seen over the entire bandwidth, with a very strong peak at 1.415 GHz. We observe RFI in this location as a consequence of the presence of two transmitters with central frequencies close to 1.428 GHz. These types of transmitter, emitting frequencies close to those of the

protected L-band, are present throughout France. A map of their locations is provided in [10]. In the case of the CEL perturbations, the RFI is limited to a small band centered around 1,406 MHz, with an approximately 3 MHz bandwidth. The brightness temperatures over the remaining parts of the spectrum appear to be close to 260 K, which should correspond to natural emission.

RFI MITIGATION ALGORITHM USING SUBBAND FILTERING

In this section we present two algorithms, designed to eliminate those parts of the spectrum corrupted by RFI. We applied in order to recover Tb values free from the influence of RFI. In order to identify the presence of RFI. An easy approach consists in dividing the spectrum into different sub-bands, and then applying different criteria to each subband, so as to classify the samples according to whether or not they are affected by RFI. The application of these algorithms requires the definition of thresholds over which samples are supposed to be corrupted. Each threshold was chosen empirically after different tests applied on the whole data set.

Algorithms Based on the 'Gaussianity' Hypothesis: Generalized Spectral Kurtosis

Kurtosis [19] is defined as the normalized form of the fourth central moment of a distribution, expressed by the ratio: $\beta_4 = \frac{\mu_4}{\mu_2^2}$ where μ_2 is the second central moment and $\mu 4$ is the fourth central moment of a statistical distribution. $\beta 4$ is equal to 3 for a Gaussian distribution (natural emission), and the deviation of this ratio from its constant value of 3 is an indicator of the presence of non-Gaussian noise (RFI). For example, the kurtosis of pulsed RFI with a duty cycle lower than 50% is greater than 3, and this property can be used to flag an RFI-contaminated subband, using a so-called generalized spectral kurtosis (SK) parameter [20,21]. This criterion can be applied to each of the sub-bands, for every radiometric sample. In the present study, we chose to eliminate data whose spectral-kurtosis was greater or lower than four times the standard deviation (STD).

A drawback of using the kurtosis parameter is that it has a blind spot for sinusoidal and chirp interference signals with a 50% duty cycle. Nevertheless, as a kurtosis algorithm was found by [22] to be the most suitable for normality analysis, we decided to test this method in our study, despite the limitation resulting from a low number of samples. For these algorithms we divided the spectrum in only eight sub-bands to evaluate the ‗gaussianity' of each sample.

Algorithm Based on the Euclidian Distance

The Euclidian distance [23] is a useful technique for determining the similarity of a sample to a set of values measured from a collection of known samples (training basis). In our application, the known samples are those considered to be uncorrupted by RFI (clean samples). The greater the Euclidian distance between the clean and the evaluated samples, the greater the likelihood of the evaluated sample is interpreted as interference noise. The Euclidian distance is defined as:

$$D_M(x) = \sqrt{(x-\mu)^T (x-\mu)} \tag{2}$$

where μ is the mean signal strength of a clean sample, and x is the evaluated sample. Once the clean samples (μ) have been empirically determined, we compute the DM values for each sub-band of all the spectra, the average (mD_M) and standard deviation (sD_M) of all D_M, and then apply a mask to the data for which the values of DM exceed $mD_M + 2\,sD_M$.

One drawback of this method (that was already proposed and tested in [14]) is the need to define one (or many if ocean survey data are present in the database) RFI-clean database, before evaluating the extent to which the remaining data is corrupted. This requirement makes the Euclidian method difficult to apply during operational procedures. It was nevertheless tested during our campaign, as it has the potential of providing a good comparative reference, which could be used to evaluate the efficiency of sub-band division.

RESULTS

The comparison of different RFI mitigation algorithms, using real data, is not a straightforward task. In the absence of an ‗*a priori*' knowledge of the data corruption level and the brightness temperature of the natural scene, we can only use indirect methods to estimate the efficiency of these algorithms. Nevertheless, one possible approach is to visually characterize this efficiency, since strong RFI can easily be identified. Additional information is provided by the quantity of data remaining, after elimination of the samples found to be RFI; a further clue is given by the mean of the final Tb value, which should be as low as possible.

Kurtosis Mask

In the case of the kurtosis algorithm, we computed the kurtosis for each subband. The sub-bands exceeding the applied threshold were not included in the brightness temperature computation. The kurtosis algorithm needs a large amount of data, to be efficient. In our study, with 8 sub-bands, we were limited to 256 values per subband, and it was not possible to increase the number of sub-bands without damage. This is a significant limitation of this criterion. The results obtained with this method are shown in Figure 5.

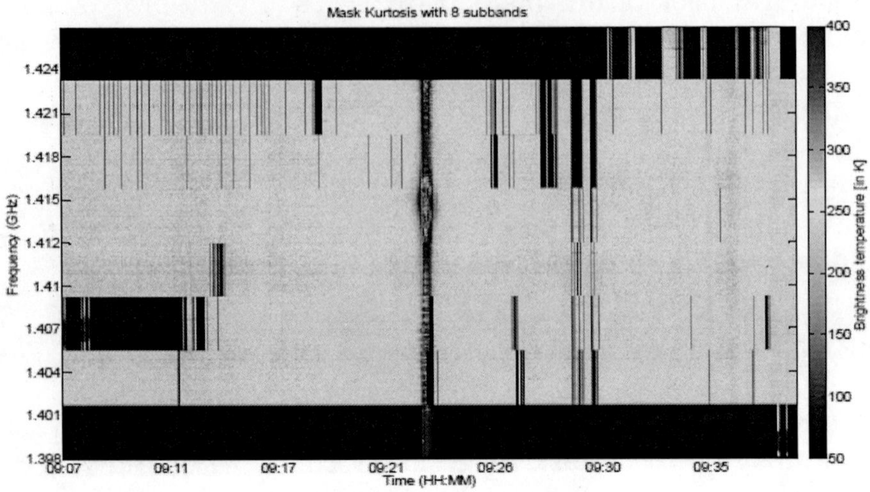

Figure 5. Spectral filtering using the kurtosis filter, and with the spectrum divided into 8 sub-bands.

The RFI present at the higher frequencies is almost always detected, as is the pulsed RFI, but it can be seen that the narrow band RFI transmitted by the CEL is not completely detected. Moreover, some parts of the spectrum collected near Auch (in the middle of the figure) are also not discriminated. Nevertheless it appears that very few false alarms are generated when using this filter, with the spectrum divided into eight sub-bands.

Euclidian Distance Mask (ED)

The influence of blanking is shown in Table 3, for several type of RFI. After correction, the signal has a level close to that corresponding to the natural microwave emission.

Table 3. Mean Tb values, before and after blanking, using the Euclidian distance, for each type of RFI.

	Natural emission	Pulsed RFI	CW RFI in 3MHz BW
Mean Tb values	267.2 K	268.2 K	279.4 K
Mean Tb values after blanking	267.2 K	267.5 K	270.0 K

Figure 7 illustrates the resulting time–frequency brightness temperature maps, before and after elimination of the corrupted sub-bands during the studied flight. The removed sub-bands are clearly seen, revealing the heterogeneity of the brightness temperature as a function of frequency. This result demonstrates the ability of the algorithm to eliminate corrupted sub-bands. For some other zones, such as the area around Auch, where a large part of the frequency band is corrupted by RFI, we did not keep the data.

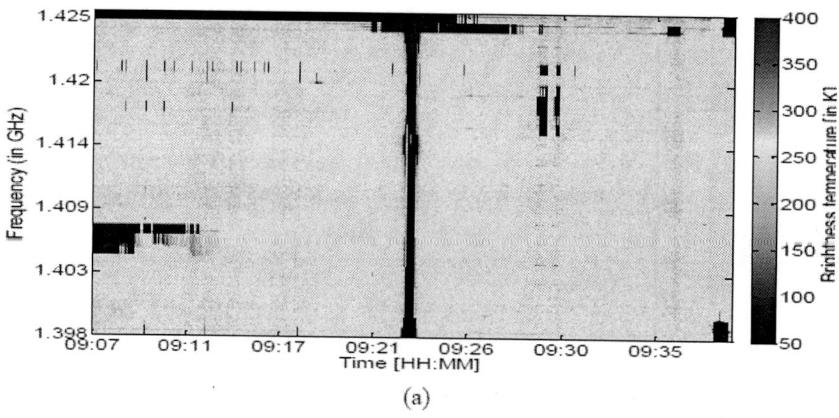

Figure 7. *Cont.*

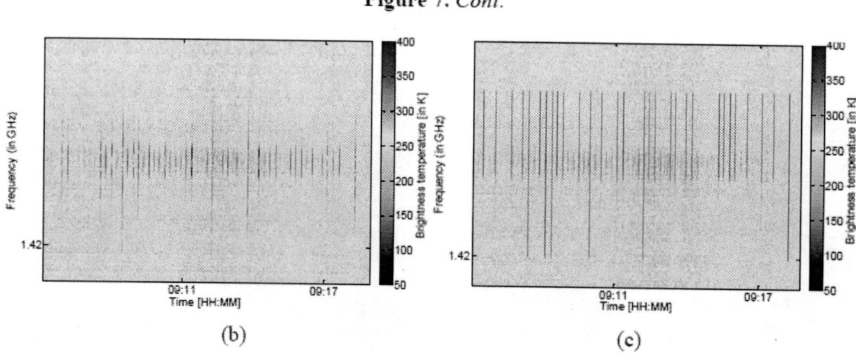

Figure 7. Brightness temperatures after correction using the Euclidian Filter, when the spectrum is divided into 32 sub-bands. The full set of data is shown in **(a)**, whereas an enlarged view of the pulsed RFI, before and after blanking, can be seen in **(b)** and **(c)** respectively.

Algorithm Comparisons

In this paragraph, we compare the two different RFI mitigation algorithms. Two factors are used in this analysis: the number of deleted data, and the number of mean Tb computed for only one small part of the transect, corresponding to a path length of approximately 40 km (close to the dimensions of the central field of view of a SMOS pixel). These values are presented in the following Table 4: the Euclidian algorithm was tested for a number of spectrum sub-bands ranging from 8 to 64.

In this table, it can be seen that mean Tb values determined for a ‗SMOS equivalent' transect vary from 270.7 K to 261.4 K, depending on the applied mask. The most efficient mask was that based on the ED criterion, using 16 sub-bands. This mask allowed only 8.3% of the data to be deleted. In that case, a low percentage of data is deleted and a low mean Tb value is reached. Because of these low mean Tb value, we can suppose that this mask deletes most of RFI (which should increase mean Tb); in the meantime, low percentage means that, compared to other masks, less false alarms are present. The kurtosis mask led to a higher mean Tb value, with a larger amount of deleted data. This outcome can be interpreted as these algorithms also mask some uncorrupted parts of the spectrum.

Table 4. % of deleted data and mean Tb values, for two different masking algorithms, and division of the spectrum into 8, 16, 32 and 64 sub-bands.

	Number of sub-bands	% of deleted data	Mean Tb value on a equivalent 'SMOS pixel'
Without Blanking			270.7
kurtosis (β_4)	8	32.2	266.3
Euclidian Distance (ED)	8	13.0	265.8
	16	8.3	261.4
	32	5.8	261.6
	64	4.0	261.9

Another way of comparing the algorithms is to compute the Tb values over the full L-band spectrum, before and after RFI masking. This comparison is made in Figure 8(a,b), which show the good accordance of these three algorithms. They all mitigate the RFI transmitted in the vicinity of the CEL site [zoom on Figure 8(b)], although the kurtosis algorithm did not efficiently mask all parts of this RFI. All the algorithms failed to eliminate the CW RFI detected near Auch, but succeeded in masking the RFI effects produced by switch-on of the on-board visible camera (around 9:28).

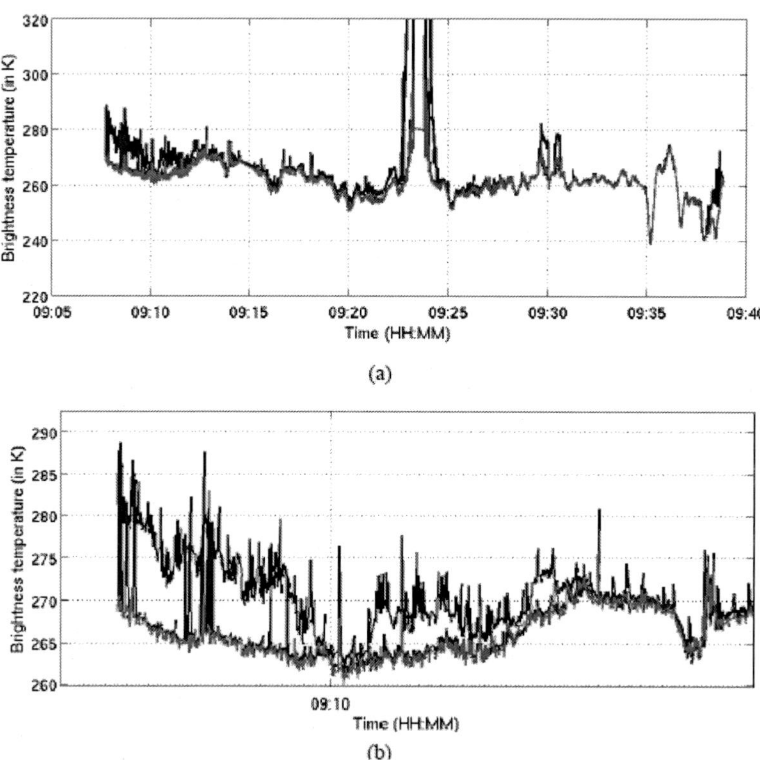

Figure 8. (a) Comparison of the different Tb values measured along the full length of the transect. **(b)** zoom on (a), near to the CEL site. The spectra are divided into 8 sub-bands in this figure. Black curve: no-blanking, blue: kurtosis, magenta: Euclidian distance.

DISCUSSION AND CONCLUSIONS

A FFT spectrum analyzer has been developed, and used to complete the French L-band CAROLS radiometer receiver, with the objective to retrieve and remove RFI from the radiometric measurements. Our study was based on airborne CAROLS measurements acquired over southwest France. Spectral analysis of the data reveals two types of RFI, the first one—such as that observed near Auch—is broadband. The second one is produced by radar pulse emissions, and is narrowband. Following RFI correction, a new —clean□ brightness temperature map, with uncorrupted sub-bands, is produced. The results from L-band observations clearly demonstrate the efficiency of the Euclidian distance algorithm, for RFI mitigation based on frequency domain blanking. Pulsed and continuous wave RFI are removed, and the brightness temperature can be correctly retrieved. This is an important result, because ocean salinity and surface soil moisture can thus be correctly estimated, even if the measurements are corrupted by RFI. We show that the other algorithm, based on kurtosis parameters, is also able to mitigate RFI on a spectral basis, with the drawback that some of the continuous waves RFI are not deleted. In the future, the Euclidian distance could be replaced by the Mahalanobis distance, calculated on the basis of polarimetric information related to subband and polarization diversity. This algorithm could improve the accuracy with which RFI and non-RFI samples are classified

ACKNOWLEDGEMENTS

This work is a SMOS/ESA Cal/Val project, and was funded by the TOSCA/CNES program. The authors would like to thank the French CNES for the considerable support it has given to the CAROLS project, as well as to all of the CAROLS campaigns. The authors would also like to thank the technical teams from LATMOS, SAFIRE, DT-INSU, DTU, SAFIRE and the pilots for their support and contributions to the successful airborne campaigns.

REFERENCES

1. 1. Njoku, E.G.; Ashcroft, P.; Chan, T.K.; Li, L. Global survey and statistics of radio-frequency interference in AMSR-E land observations. *IEEE Trans. Geosci. Remote Sens.* **2005**, *43*, 938-947.

2. 2. Kerr, Y.H.; Waldteufel, P.; Wigneron, J.-P.; Martinuzzi, J.-M, Font, J., Berger, M. Soil moisture retrieval from space: The soil moisture and ocean salinity (SMOS) mission. *IEEE Trans. Geosci. Remote Sens.* **2001**, *39*, 1729-1736.

3. 3. Entekhabi, D.; Njoku, E.G.; O'Neill, P.E.; Kellogg, K.H.; Crow, W.T.; Edelstein, W.N.; Entin, J.K.; Goodman, S.D.; Jackson, T.J.; Johnson, J.; Kimball, J.; Piepmeier, J.R.; Koster, R.D.; Martin, N.; McDonald, K.C.; Moghaddam, M.; Moran, S.; Reichle, R.; Shi, J.C.; Spencer, M.W.; Thurman, S.W.; Leung, T.; Van Zyl, J. The Soil Moisture Active Passive (SMAP) Mission. *Proc. IEEE* **2010**, *98*, 704-716.

4. 4. Fridman, P.A.; Baan, W.A. RFI mitigation methods in radio astronomy. *Astron. Astrophys.* **2001**, *378*, 327-344.

5. 5. Johnson, J.T.; Gasiewski, A.J.; Güner, B.; Hampson, G.A.; Ellingson, S.W.; Krishnamachari, R.; Niamsuwan, N.; McIntyre, E.; Klein, M.; Leuski, V.Y. Airborne radio-frequency interference studies at C-band using a digital receiver. *IEEE Trans. Geosci. Remote Sens.* **2006**, *44*, 1974-1985.

6. 6. Ruf, C.S.; Gross, S.M.; Misra, S. RFI detection and mitigation for microwave radiometry with an agile digital detector. *IEEE Trans. Geosci. Remote Sens.* **2006**, *44*, 694-706.

7. 7. Piepmeier, J.R.; Mohammed, P.N.; Knuble, J.J. A double detector for RFI mitigation in microwave radiometers. *IEEE Trans. Geosci. Remote Sens.* **2008**, 46, 458-464.

8. 8. Niamsuwan, N.; Johnson, J.T.; Ellingson, S.W. Examination of a simple pulse blanking technique for RFI mitigation. *Radio Sci.* **2005**, *40*, 11.

Chapter 8

SIMULTANEOUS MEASUREMENTS OF CHLOROPHYLL CONCENTRATION BY LIDAR, FLUOROMETRY, ABOVE-WATER RADIOMETRY, AND OCEAN COLOR MODIS IMAGES IN THE SOUTHWESTERN ATLANTIC

Milton Kampel [1,*], João A. Lorenzzetti [1], Cristina M. Bentz [2], Raul A. Nunes [3], Rodolfo Paranhos [4], Frederico M. Rudorff [1] and Alexandre T. Politano [2]

[1]Instituto Nacional de Pesquisas Espaciais (INPE), PO Box 515, 12201-970, São José dos Campos, SP, Brazil; E-mails: {loren, fmr}@dsr.inpe.br

[2]PETROBRAS/CENPES, Cidade Universitária, Q.7, Ilha do Fundão, 21949-900, Rio de Janeiro, RJ, Brazil; E-mails: {cris, politano.gorceix}@petrobras.com.br

[3]Pontifícia Universidade Católica do Rio de Janeiro (PUC-Rio), PO Box 38008, 22453-900, Rio de Janeiro, RJ, Brazil; E-mail: nunes@dcmm.puc-rio.br

[4]Universidade Federal do Rio de Janeiro (UFRJ), Cidade Universitária, Av. Pau Brasil 211, Ilha do Fundão, 21941-590, Rio de Janeiro, RJ, Brazil; E-mail: rodpar@biologia.ufrj.br

* Author to whom correspondence should be addressed; E-Mail: milton@dsr.inpe.br; Tel.: +55 12 39456477; Fax: +55 12 39456488

ABSTRACT

Comparisons between *in situ* measurements of surface chlorophyll-*a* concentration (CHL) and ocean color remote sensing estimates were conducted during an oceanographic cruise on the Brazilian Southeastern continental shelf and slope, Southwestern South Atlantic. *In situ* values were based on fluorometry, above-water radiometry and lidar fluorosensor. Three empirical algorithms were used to estimate CHL from radiometric measurements: Ocean Chlorophyll 3 bands (OC3MRAD), Ocean Chlorophyll 4 bands (OC4v4RAD), and Ocean Chlorophyll 2 bands (OC2v4RAD). The satellite estimates of CHL were derived from data collected by the MODerate-resolution Imaging Spectroradiometer (MODIS) with a nominal 1.1 km resolution at nadir. Three algorithms were used to estimate chlorophyll concentrations from MODIS data: one empirical - OC3MSAT, and two semi-analytical - Garver, Siegel, Maritorena version 01 (GSM01SAT), and CarderSAT. In the present work, MODIS, lidar and *in situ* above-water radiometry and fluorometry are briefly described and the estimated values of chlorophyll retrieved by these techniques are compared. The chlorophyll concentration in the study area was in the range 0.01 to 0.2 mg/m3. In general, the empirical algorithms applied to the *in situ* radiometric and satellite data showed a tendency to overestimate CHL with a mean difference between estimated and measured values of as much as 0.17 mg/m3 (OC2v4RAD).

The semi-analytical GSM01 algorithm applied to MODIS data performed better (*rmse* 0.28, *rmse-L* 0.08, *mean diff.* -0.01 mg/m3) than the Carder and the empirical OC3M algorithms (*rmse* 1.14 and 0.36, *rmse-L* 0.34 and 0.11, *mean diff.* 0.17 and 0.02 mg/m3, respectively).

We find that *rmsd* values between MODIS relative to the *in situ* radiometric measurements are < 26%, *i.e.*, there is a trend towards overestimation of *RRS* by MODIS for the stations considered in this work. Other authors have already reported over and under estimation of MODIS remotely sensed reflectance due to several errors in the bio-optical algorithm performance, in the satellite sensor calibration, and in the atmospheric-correction algorithm.

INTRODUCTION

Ocean-color remote sensing has changed our perspective of ocean observation. Global maps of surface chlorophyll concentration (CHL), a proxy of phytoplankton primary productivity can be routinely produced [1, 2]. However, these satellite products are based on the application of generally complex bio-optical algorithms [3], including atmospheric correction models [4] applied on the radiance values measured by the remote sensors to estimate the water leaving radiances. In order to guarantee good quality and long term consistency of satellite data time series, orbital images must be calibrated and validated with the use of in situ measurements from research ships, moored buoys and drifters, for example [3].

Satellite images have been systematically used for monitoring the oceanic environment on the Brazilian Southeastern continental margin. However, only very seldom, has simultaneous in situ data been available in this region for the assessment of satellite algorithm accuracies. Aiming to advance scientifically and technically the current remote sensing data analysis procedures a research project -FITOSAT was jointly conducted by INPE, PETROBRAS R&D Center and other universities. One of the phases of this project involved an oceanographic cruise with the simultaneous acquisition of in situ and remote sensing data in the Campos Basin region. The simultaneous data acquisition enabled the meteo-oceanographic contextualization of the in situ data collection and allowed the evaluation of the remote sensing products.

Among other measurements, in situ estimates of CHL were obtained with laboratory fluorometry, above-water radiometry and through a lidar fluorosensor. The satellite estimates of CHL were derived from MODIS data. In this article, these different methods used for estimating CHL are briefly described and their results are statistically compared.

Data and Methods

In situ Chlorophyll

Chlorophyll measurements were done at 18 stations during an oceanographic cruise held in November 2004. The study area included the Brazilian Southeastern continental shelf and slope region, from Cape Sao Tome (22oS) to Cape Frio (23oS), at Campos Basin, Rio de Janeiro (Figure 1).

Phytoplankton abundance was estimated from 2 L surface water samples filtered through Millipore cellulose membranes (0.45 µm). The filters were kept in liquid nitrogen and the chlorophyll-a concentrations were determined after extraction in 90% acetone for 18 hours at 4oC in a Turner TD- 700 fluorometer [5].

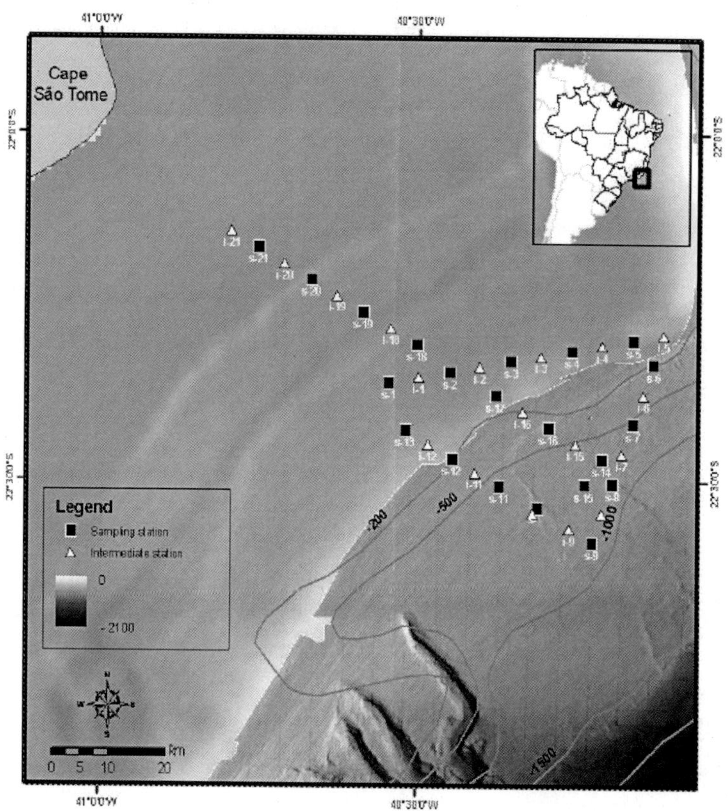

Figure 1. Study area and field stations occupied during November 2004 at the Brazilian Southeastern continental margin. Water sampling, above-water radiometry and lidar stations labeled with an "s" (black squares). Stations without water sampling labeled with an "i" (white triangles). Isobaths in meters.

In situ Radiometry

Above water radiometric measurements of water leaving radiance and incident irradiance were obtained with a hyperspectral SPECTRON SE590 radiometer at 30 stations (Figure 1). Each radiance spectrum was sampled between 400–800 nm, with a 5 nm resolution. The protocol adopted for the measurements was the one proposed by Fougnie et al. [6], with a polarizer filter. The radiometric data were numerically integrated to simulate the spectral bands of

SeaWiFS and MODIS orbital sensors, using the trapezoidal rule. The remote sensing reflectance, RRS(λ), was calculated by the following equation:

$$R_{RS} = \frac{L_w(\lambda)}{E_d(\lambda)} \tag{1}$$

where Lw(λ) is the water leaving spectral radiance and Ed (λ) is the downwelling spectral irradiance incident on the sea surface. Ed (λ) was estimated by the radiance reflected by a Spectralon plate [7], as follows:

$$E_d(\lambda) = L(\lambda) f_c \pi \tag{2}$$

where fc is a correction factor estimated in the laboratory by the ratio of Lref (λ) of a standard reference (approximately 100%) by the Spectralon plate's L(λ) used in the field. The SeaWiFS empirical algorithms Ocean Chlorophyll 4-bands - OC4v4 and Ocean Chlorophyll 2-bands - OC2v4, and the MODIS algorithm Ocean Chlorophyll 3-bands - OC3M were applied to the radiometric data to estimate CHL.

The OC2v4 algorithm estimates CHL based on a band ratio of RRS(490)/RRS(555) using a modified cubic polynomial function [3]:

$$CHL = 10{,}0^{(0{,}319 - 2{,}336 R_{2S} + 0{,}879 R_{2S}^2 - 0{,}135 R_{2S}^3) - 0{,}071} \tag{3}$$

Where $R_{2S} = \log_{10}\left(R_{555}^{490}\right)$

The algorithm OC4v4 also relates a band ratio with CHL using a polynomial function, but is based on the maximum band ratio determined as the highest ratio (Rmax) between the values of RRS(443)/RRS(555), RRS(490)/RRS(555), and RRS(510)/RRS(555) through the following function of forth order [3]:

$$CHL = 10{,}0^{(0{,}366 - 3{,}067 R_{4S} + 1{,}930 R_{4S}^2 - 0{,}649 R_{4S}^3 - 1{,}532 R_{4S}^4)} \tag{4}$$

Where $R_{4S} = \log_{10}(R_{\max})$

The algorithm OC3M also uses a polynomial function of forth order from Rmax between RRS(443)/RRS(550) and RRS(490)/RRS(550), through the following equation [3]:

$$CHL = 10,0^{(0,283-2,753R_{3M}+1,457R_{3M}^2-0,659R_{3M}^3-1,403R_{3M}^4)} \qquad (5)$$

Where $R_{3M} = \log_{10}(R_{\max})$

Lidar

The PUC-Rio fluorosensor lidar system operated in a pulse mode uses the excitation line at 532 nm of the doubled frequency radiation (second harmonic) of Q-switch Nd:YAG laser as light source. A rotating rectangular metal mirror directs the probing beam to the water surface and deflects the water backscattered signal to a 200 mm-diameter Newton-type reflecting telescope, which is filtered in order to eliminate elastic backscattering. This return signal passes through the entrance slit of the polychromator, which is assembled as an auto collimation lens focusing system. A 1024 elements array CCD camera with Peltier cooling system was used to detect the spectrum of the return signal.

Two types of radiation are collected: one is the water-Raman backscattering (which occurs at 655 nm) and the other is the fluorescence radiation generated when the light source interacts with the fluorescent elements present in the water. The laser light at this wavelength excites the fluorescence of chlorophyll, in a band centered at 685 nm, and of dissolved organic matter, in a region from 540 to 620 nm. Over 13.000 spectra were obtained during the campaign, with the lidar installed at the gangway of the R/V Astro-Garoupa. In order to validate each spectrum, an analysis of the Fourier Transform power spectrum was made, and then a validation criterion of these spectra was established. Of the total number of spectra collected during the vessel's course between the November

21st and 25th, 9,511 of them were the objects of analysis.

The intensities of the chlorophyll bands (Icl) were calculated according to the method developed by Barbosa [8] for the equipment. The values of CHL were obtained by means of an adjustment using calibration parameters obtained via laboratorial analysis of the samples collected during the present cruise and from a prior cruise at the same region. The calibration used is given by:

$$CHL_{LIDAR} = 4{,}9 * (I_{cl}) - 0{,}175 \qquad (6)$$

Remote Sensing

Ocean color remote sensing images acquired by MODIS sensor during the same period of the oceanographic cruise were processed as CHL fields with the application of one empirical and two semi-analytical algorithms. The MODIS sensor has 36 spectral bands, with eight dedicated to ocean color applications. These bands have a spatial nominal resolution of to 1.1 km and a temporal resolution of 1-2 days.

MODIS images were acquired locally by INPE's receiving station and processed using SeaDAS software distributed by NASA. Initially, the data were radiometrically calibrated to generate the water normalized upwelling radiances. The images considered of interest were selected in accordance to the study area. Atmospheric correction algorithms [4] were applied to each image using the more updated atmospheric ancillary files available before the calculation of CHL values.

As mentioned before, the OC3M empirical algorithm [3], and two semi-analytical algorithms, Garver, Siegel, Maritorena version 01 - GSM01 [9], and Carder [10] were applied to estimate CHL with MODIS data. Maritorena et al. [9] presented a protocol to improve the semi-analytical model initially proposed by Garver and Siegel [11], for global applications. The complete formulation of the model can be expressed as the following:

$$L_{WN}(\lambda)=\frac{tF_0(\lambda)}{n_w^2}\sum_{i=1}^{2}g_i\left\{\frac{b_{bw}(\lambda)+b_{bp}(\lambda_0)(\lambda/\lambda_0)^{-\eta}}{b_{bw}(\lambda)+b_{bp}(\lambda_0)(\lambda/\lambda_0)^{-\eta}+a_w(\lambda)+Chl^{*}_{ph}(\lambda)+a_{cdm}(\lambda_0)\exp[-S(\lambda-\lambda_0)]}\right\} \quad (7)$$

where L_{WN} is the normalized water leaving radiance, t is the air-sea transmission factor; F0(λ) is the extra-terrestrial solar irradiance; n_w is the refraction index of water; $g_1 = 0.0949\ sr^{-1}$ and $g_2 = 0.0794\ sr^{-1}$; $bb_w(\lambda)$ is the backscattering of water; $a_w(\lambda)$ is the absorption by water; $bb_p(\lambda)$ is the backscattering by particles; Chl_{ph}^{*} is the chlorophyll-a specific absorption coefficient; S is the spectral decay for the dissolved matter and detritus absorption (cdm); η is the exponent of the power law for the particulate backscattering coefficient; λ 0 is the wavelength 443 nm wavelength.

The Carder algorithm [10] utilizes a more complex approach. The components associated with the absorption of the pigments are divided from those associated with the degradation products (for example, gelbstoff and detritus, $a^{*}_{g}(\lambda)$). The absorption coefficient of phytoplankton chlorophyll, $a^{*}_{ph}(\lambda)$ is adjusted in relation to the chlorophyll concentration and the availability of light and nutrients. The distinction between the effects of the principal constituents is obtained by the spectral differences between $a^{*}_{g}(\lambda)$ and $a^{*}_{ph}(\lambda)$. Comparing sea surface temperature with nitrate depletion temperature (NDT) [12], the presence of big cells rich in chlorophyll and small cells poor in chlorophyll can be deduced from the satellite data [10]. The chlorophyll rich cells with low values of $a^{*}_{ph}(\lambda)$, i.e., with packed pigments, occurs generally in ambient with low level of light and rich in nutrients. On the other hand, chlorophyll poor cells but with high values of $a^{*}_{ph}(\lambda)$, i.e., without packaging, are present in ambient replete of photons, but poor in nutrients. During the development and validation of the Carder algorithm, in situ data sets were compartmented into 2 regions. In the first, the pigment packaging would not be expected, and a second one where this packaging effect would probably occur more frequently or more intensely. Besides, a global average algorithm was developed to be used in conditions where the packaging effect is unknown or transitional. The algorithm also alternates between the empirical and the semi-analytical formulation, using different coefficients for the varying levels of pigment packaging.

RESULTS AND DISCUSSION

Statistical comparisons

The comparisons obtained between the in situ CHL and those estimated with MODIS data were calculated inside a time window of 12 hours. The pairs of data were composed between the CHL insitu and the median value of a 3 x 3 pixels (9 km^2) centered on the geographical position of the sampling station in the equivalent MODIS image. The fluorometric data were statistically compared to the satellite estimates, above-water radiometric data and lidar data, through linear regression analysis, root mean square error (rmse), and transformed-rmse (rmse^{-L}) [10].

CHL insitu values varied between 0.077 and 0.197 mg/m^3 with a mean value of 0.12 (±0.04) mg/m^3. These low values are typical of the oligotrophic waters of the Brazil Current (BC) as observed previously by other authors [13, 14]. With Z90 depths [15] of ~13 m, the chlorophyll concentration was constant over one optical depth in 8 stations, with an increase in 10 stations. The mean chlorophyll concentration to a depth of one optical depth was 0.15 (±0.04) mg/m3, with a maximum value of 0.303 mg/m^3.

In general, the empirical algorithms applied to the in situ radiometric data overestimated CHL insitu (Figure 2). OC3M and OC4v4 presented a similar performance, with rmse-L (1.09, 1.08) and rmse (0.33, 0.32) as compared to OC2v4 (1.57 and 0.45), respectively (see Table 1). But the lower mean difference between estimated and measured values was calculated for OC4v4 (0.01 mg/m3).

In addition to the bio-optical algorithm performance, there may be some errors in the satellite sensor calibration and in the atmospheric-correction algorithm (see below).

Based on the MODIS chlorophyll fields, in situ data and numerical modeling was possible to observe an anticyclonic Brazil Current frontal eddy in the study area during the cruise [16]. Concurrent with the mesoscale eddy there was an uncommon sea floor oil seep event [17], which surface signature won't be discussed in detail here.

For illustration, examples of MODIS images processed using with the three algorithms tested in this study are shown (Figure 3). In the

11/25/2004 image, the oligotrophic waters of the BC are observed offshore, over the slope region, in dark blue colors (OC3M<0.08 mg/m^3; GSM01<0.08 mg/m^3; Carder<0.10 mg/m^3). A surface signature of the mesoscale eddy with a mean diameter of 75-80 km was identified in front of Cape Sao Tome between 22.00oS–22.65oS, and 39.9oW–40.77oW, with relatively higher CHL values (0.10<OC3M<0.15 mg/m^3; 0.08<GSM01<0.13 mg/m^3; 0.20<Carder<0.30 mg/m^3).

In order to perform a quantitative analysis of differences between the images produced by the three algorithms, we compared the histograms of the full area (not shown) and of the in situ sampled area (Figure 4, Table 2). Considering the full area, the CHL values varied similarly (0.011–19.929 mg/m3 OC3M, 0.010–19.862 mg/m3 Carder, 0.010–19.154 mg/m3 GSM01). But the mean value of the GSM01 (0.15 ± 0.25 mg/m3) was relatively lower than OC3M (0.45 ± 1.26 mg/m3) and Carder (0.43 ± 1.07 mg/m3). Now considering the in situ sampled area (Figure 4) the relative pattern was the same (Table 2) with the GSM01 mean value (0.10 ± 0.07 mg/m3) relatively lower than OC3M (0.21 ± 0.27 mg/m3) and Carder (0.23 ± 0.27 mg/m3).

Table 1. Comparison between different in situ and satellite estimates of surface chlorophyll-a concentration in the Brazilian Southeastern continental shelf and slope waters in November 2004. (N= number of observations; RAD= in situ radiometry; SAT= MODIS data).

Ocean color satellite algorithms OC3M and Carder overestimated on average the values of CHL insitu, with a mean difference of 0.02 and 0.17 mg/m3, respectively (Table 1), but the algorithm GSM01 has underestimated the relatively higher in situ concentrations (Figure 2). Statistically, the best performance was obtained with the semi-analytical algorithm GSM01, with the lowest values of rmse-L (0.28), rmse (0.08), and mean difference (-0.01 mg/m3), respectively. The empirical algorithm OC3M also presented a good performance, with rmse-L equal to 0.36, rmse of 0.11, and mean difference of 0.02 mg/m3 (see Table 1).

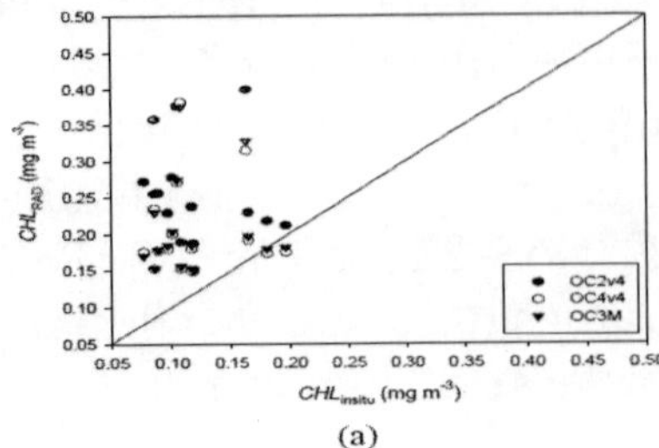

(a)

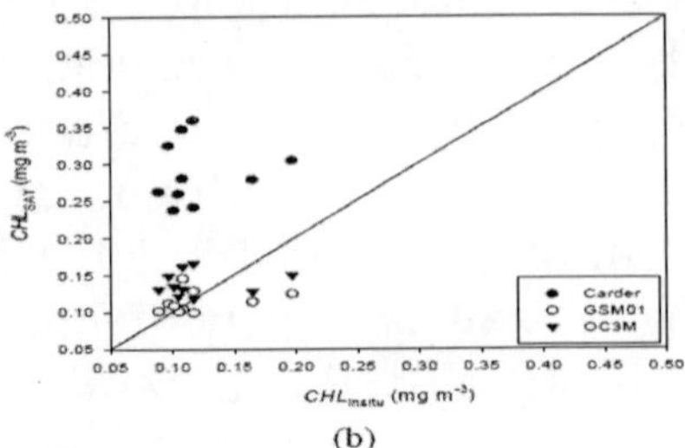

(b)

Figure 2. Scatter plots of chlorophyll concentration estimates obtained with (a) above water radiometric data (three algorithms), and (b) MODIS ocean color remote sensing data (three algorithms) against in situ fluorometric estimates.

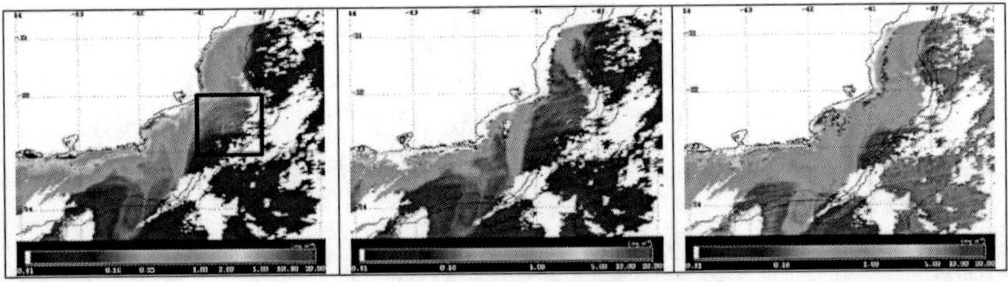

Figure 3. Surface distributions of chlorophyll concentration estimated with OC3M (left), GSM01 (center), and Carder (right) algorithms applied to MODIS data acquired on 11/25/2004 (see text for details). Isobaths in meters. Color table in logarithmic scale. Land and clouds are masked in white. Black square on the left indicates the in situ sampled area. There are some discrepancies between the datasets, which may be due to different reasons. A relevant aspect is that there is usually a time difference between the satellite and in situ measurements. This could be particularly significant when the chlorophyll distribution features are dynamic, as during the cruise period when an anticyclonic eddy was active in the study area. Besides, to evaluate semianalytical algorithms properly, inherent optical properties measurements are required, but these measurements were not acquired for this work. The absorption coefficient is one of the inherent optical properties that affect the reflectance of the aquatic systems. Therefore, the knowledge about the total absorption and its components is fundamental to improve the description of the spectral reflectance variability and chlorophyll concentration estimated by ocean color remote sensing sensors [18].

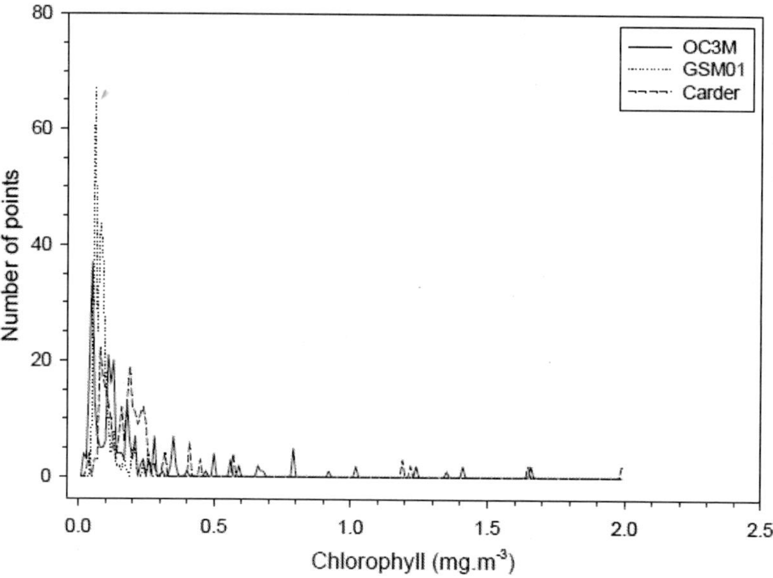

Figure 4. Histograms of MODIS chlorophyll concentration estimates obtained with OC3M, GSM01, and Carder algorithms for the in situ sampled area on 11/25/2004 (see text for details).

Table 2. Comparison between histograms of MODIS chlorophyll concentration estimates obtained with 3 algorithms for the full image area and for the in situ sampled area (mg/m^3).

Algorithm	*Min.*	*Max.*	*Mean*	*std.*
Full area				
OC3M	0.01	19.93	0.45	1.26
GSM01	0.01	19.15	0.15	0.25
Carder	0.01	19.86	0.43	1.07
Sampled area				
OC3M	0.02	1.66	0.21	0.27
GSM01	0.05	0.26	0.10	0.04
Carder	0.03	2.00	0.23	0.27

The spectral distributions of RRS measured at 30 stations are shown on Figure 5. In general, all the spectra are typical of Case 1 oceanic and oligotrophic waters with low chlorophyll concentrations. The noise observed at station S-5 (red on Figure 5) was due to the low light level at the time of data acquisition. The different shape of the spectrum at station S-21 corresponds to more coastal waters (green on Figure 5), with relatively higher chlorophyll concentration (0.54 mg/m^3 OC3M) and with the probable presence of other optically active components.

A spectral comparison between average MODIS-derived RRS and the in situ estimates from the radiometer for 15 "match-up" stations is shown in Figure 6. The 1:1 line of perfect correspondence is also shown. We define rmsd at any wavelength λ as [19]:

$$rmsd(\lambda) = 100\sqrt{\left[N^{-1}\sum\frac{(R_{RS}^{SAT} - R_{RS}^{RAD})^2}{(R_{RS}^{RAD})}\right]}, \quad (8)$$

where SAT RS R and RAD RS R are the remote sensing reflectance estimates from MODIS and the above-water radiometer, respectively at the same station and N is the number of matched stations, where the summation Σ is over N stations (here N = 15). We find that rmsd values between MODIS relative to the in situ radiometric measurements are 25.3% at 412 nm, 25.9% at 443 nm, 25.9% at 488 nm, 15.6 % at 531 nm, and 13.2% at 551 nm, i.e., largest for 443–448 nm, and least for 551 nm.

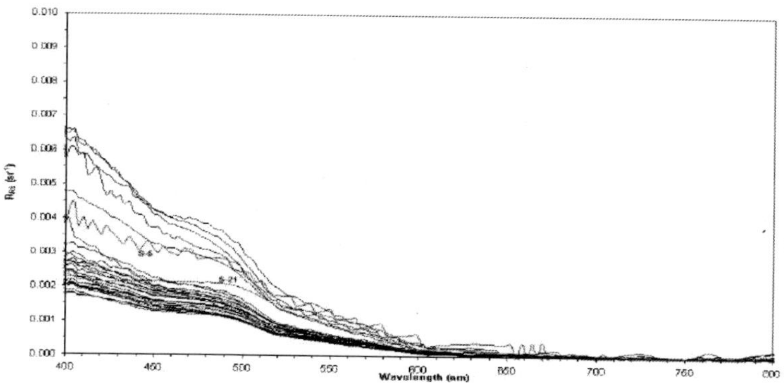

Figure 5. Surface remote sensing reflectance spectra obtained by above-water radiometry during the cruise FITOSAT I cruise, November 2004, in the Brazilian Southeastern continental shelf and slope. Stations S-5 and S-21 in red and green, respectively (see text).

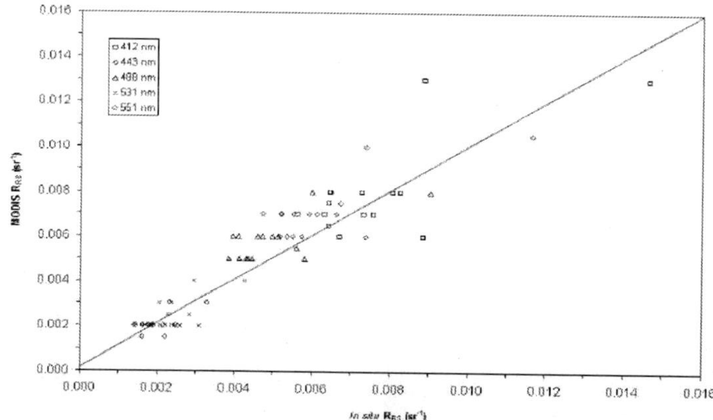

Figure 6. Scatter plot of MODIS remote sensing reflectance (RRS) with respect to in situ above-water RRS for 15 stations where inter-comparison were made.

It was shown that RRS measured RRS and estimated from MODIS are generally in agreement. The OC3M empirical algorithm overestimated CHLinsitu when applied to measured RRS and to satellite derived RRS. However, OC3M bio-optical algorithm performed much better with MODIS-derived RRS (see Table 1). The reduced number of satellite "match-ups" available for analysis (N =

10) compared to the in situ radiometric measurements (N = 17) may partially account for the difference found.

Furthermore, as mentioned before, there are some discrepancies between the datasets, which may be due to different reasons. The vertical chlorophyll profile was homogeneous over one optical depth only in 8 stations, with concentration increasing in other 10 stations. And the "match-up" stations are not coincident. In addition, we speculate here that compensation could occur between errors in atmospheric correction, satellite sensor calibration and uncertainty in the bio-optical algorithm.

The fluorescence spectra processed from the lidar data are presented on Figure 7. The chlorophyll concentrations estimated with the lidar (CHLLIDAR) at the stations with simultaneous water sampling for fluorometric analysis varied between 0.064 and 0.016 mg/m3, with a mean value of 0.12 (±0.03)mg/m3, very similar to CHL insitu. In fact, a paired t-test confirmed that the two mean values are statistically equal. The performance of the lidar was similar to that obtained by the in situ fluorometric and radiometric methods, with rmse, rmse-L and mena diff. values of 0.14, 0.48, and 0.01 mg/m3 respectively.

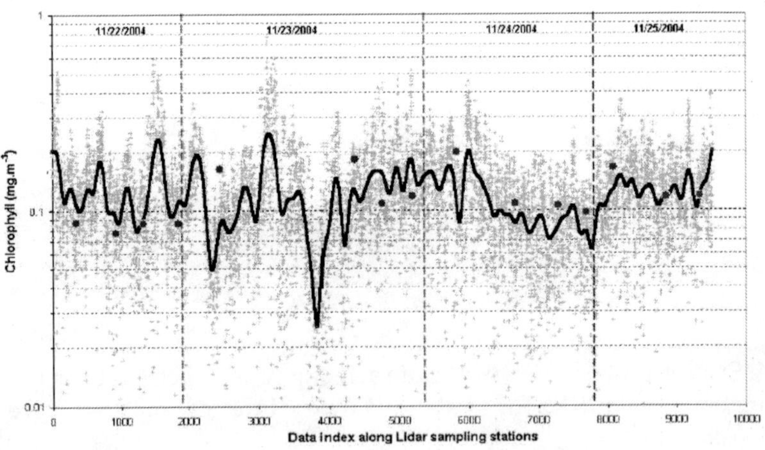

Figure 7. Surface chlorophyll concentration estimated with the along-track lidar sampling during the cruise FITOSAT I cruise, November 2004, in the Brazilian Southeastern continental shelf and slope compared with fluorometric

estimates (grey crosses - Lidar measurements; black line - Lidar processed; red circles - fluorometric).

The amplitude of laser excited fluorescence on Figure 7, where the observations were conducted at the same point, can be associated with a diurnal variation. The laser excited fluorescence exhibits a high fluorescence in the night while phytoplankton is not conducting photosynthesis. During the day, exhibited a low fluorescence while phytoplankton is already excited by the solar irradiation and emitting less fluorescence against the laser excitation. The relationship between phytoplankton biomass and pigment content varies widely with light acclimation and nutrient status, while the fluorescence yield also varies with the physiological condition of the phytoplankton and with the type of plankton present. In both laboratory studies and in the field, researchers have observed a daily rhythm of fluorescence that is not correlated with diel changes in the concentration of chlorophyll-a. During periods of high irradiance fluorescence tends to be lower than it is at night [20–22].

We learned from previous experiments that our lidar measurements are well correlated with the surface (~1 m) chlorophyll concentration. The analysis conducted to adjust the lidar with the fluorometric estimates was based on the observation that de relative standard deviation (variance coefficient) of the latter was much higher than the former. As mentioned on Section 2.3, the values of CHLLIDAR were obtained by means of a single equation (see Equation 6).

Spectrally resolved local and remote laser induced fluorescence detectors are emerging as very reliable tools for the collection of information on fluorescent targets for sea and land diagnostics purposes [23, 24]. Barbini et al. [25] also compared lidar data and local spectrophotometer and/or spectro-fluorometer measurements in Antarctica. The observed discrepancies were explained as a result of the fact that lidar signals were collected from a layer where mixing processes can take place, whereas seawater samples were collected at a well defined depth.

Barbini et al. [26] compared lidar measurements of surface chlorophyll-a concentrations in the Ross Sea and in the Antarctica-

New Zealand transect to data collected by SeaWiFS. The authors concluded that in coastal waters, lidar values were significantly higher than the orbital estimates. In open waters the differences were smaller and the sampling stations were considered robust to calibrate the SeaWiFS algorithm against the lidar measurements in the Ross Sea.

CONCLUSIONS

To our best knowledge this is the first published result of simultaneous measurements of surface chlorophyll concentration in the Brazilian Southeastern continental shelf and slope region using in situ fluorometric, above-water radiometry, and lidar, complemented with ocean color remote sensing MODIS imagery.

The empirical algorithms applied to the in situ radiometric and satellite data showed a tendency of overestimating the in situ CHL values by as much as 0.17 mg/m3 (for OC2v4RAD). The semianalytical GSM01 algorithm applied to MODIS data presented the best performance (rmse 0.28, rmse-L 0.08, mean diff. -0.01 mg/m3). With the available dataset, MODIS showed a trend in overestimating the RRS (up to 26%) relative to the in situ measurements.

When a comprehensive bio-optical dataset from future cruises is collected, a regional algorithm can be developed, to suit local Brazilian Southeastern continental shelf and slope waters. However, this new algorithm must be able to estimate not only CHL, but also other optically active constituents to allow the monitoring of other important water quality parameters and to advance in the ecosystemic study of the interest area.

A better calibration of the lidar equipment is necessary to account for the daily rhythm of fluorescence that is not correlated with diel changes in CHL. This would provide the opportunity to have a CHL onboard monitoring equipment that would automatically and autonomously determine the surface chlorophyll concentrations along the ship track.

ACKNOWLEDGEMENTS

This work was supported by the PETROBRAS R&D Center, the National Space Research Institute (INPE), and institutions involved in the FITOSAT Project, including the Federal University of Rio de Janeiro and the Catholic University of Rio de Janeiro. NASA provided the satellite images and SeaDAS software. The authors would like to thanks the crew of R/V Astro Garoupa, researchers and students that participated in the field work, and for the comments of two anonymous referees.

REFERENCES AND NOTES

1. 1. Falkowski, P.G.; Raven, J.A. Aquatic Photosynthesis. Blackwell: Malden, Massachusetts, 1997; p. 37.
2. 2. Behrenfeld, M.J.; Falkowski, P.G. Photosynthetic rates derived from satellite-based chlorophyll concentration. Limn. Oceanogr. 1997, 42, 1–20.
3. 3. O'Reilley, J.E.; Maritorena, S.; O'Brien, M.C.; Siegel, D.A.; Toole, D.; Menzies, D.; Smith, R.C.; Mueller, J.L.; Mitchell, B.G.; Kahru, M.; Chavez, R.P.; Strutton, P.; Cota, G.F.; Hooker, S.B.; McClain, C.R.; Carder, K.L.; Mueller-Karger, F.; Harding, L.; Magnusion, A.; Phynney, D.; Moore, G.F.; Aiken, J.; Arrigo, K.R.; Letelier, R.; Culver, M. In: SeaWiFS Postlaunch Calibration and Validation Analyses, Part 3, Volume 11. S.B. Hooker and E.R. Firestone (Ed.); NASA/TM-2000-2206892. NASA Goddard Space Flight Center: Greenbelt, MD, 2000.
4. 4. Gordon, H.R.; Wang, M. Retrieval of water-leaving radiance and aerosol optical thickness over the oceans with SeaWiFS: a preliminary algorithm. Appl. Opt. 1994, 33, 443–452.
5. 5. Parsons, T.R.; Maita, Y.; Lalli, C.M. A manual of chemical and biological methods for seawater analysis. Pergamon Press: Oxford, 1984; p. 173.
6. 6. Fougnie, B.; Frouin, R.; Lecomte, P.; Deschamps, P.-Y. Reduction of skylight reflection effects in the above-water measurement of diffuse marine reflectance. Appl. Opt. 1999, 38, 3844–3856.
7. 7. Mobley, C.D. Estimation of the remote-sensing reflectance from above-surface measurements. App. Opt. 1999, 38, 7442–7455.
8. 8. Barbosa, P.C.C. Aplicações de Fluorescência Induzida por Laser em Monitoramento Ambiental. (PUC-Rio, No 9916186/CA). PhD Thesis. Pontifícia Universidade Católica do Rio de Janeiro: Rio de Janeiro, Brazil, 2003 (in Portuguese).
9. 9. Maritorena, S.; Siegel, D.A.; Peterson, A.R. Optimization of a semi-analytical ocean color model for global-scale applications. Appl. Opt. 2002, 41, 2705–2714.

10. 10. Carder, K.L.; Chen, F.R.; Cannizzaro, J.P.; Campbell, J.W.; Mitchell, B.G. Performance of the MODIS semi-analytical ocean color algorithm for chlorophyll-a. Adv. Space Res. 2004, 33, 1152–1159.

11. 11. Garver, S.A.; Siegel, D.A. Inherent optical property inversion of ocean color spectra and its biogeochemical interpretation. I. Time series from the Saragasso Sea. J. Geophys. Res. 1997, 102, 18607–18625.

12. 12. Kamykowski, D. A preliminary biophysical model of the relationship between temperature and plant nutrients in the upper ocean. Deep-Sea Res. I 1987, 34, 1067–1079.

13. 13. Kampel, M. Estimativa da produção primária e biomassa fitoplanctônica através de sensoriamento remoto da cor do oceano e dados in situ na costa sudeste brasileira. PhD Thesis, Universidade de São Paulo, Instituto Oceanográfico, São Paulo, Brazil, 2003.

14. 14. Gaeta, S.A.; Metzler, P.M.; Lopes, R.M. Plâncton e produtividade primária na plataforma da costa norte do Estado do Rio de Janeiro-Campanha de verão (dez/91-jan/92). Relatório do Projeto Monitoramento Oceânico da Bacia de Campos FUNDESPA/GEOMAP-PETROBRÁS, 1993.

15. 15. Smith, R.C. Remote sensing and depth distribution of ocean chlorophyll. Mar. Ecol. Prog. Ser. 1981, 5, 359–361.

16. 16. Lorenzzetti, J.A.; Kampel, M.; Bentz, C.M.; Torres, A.R., Jr. A meso-scale Brazil Current frontal eddy: observations by ASAR, Radarsat-1, complemented with visible and infrared sensors, in situ data, and numerical modelling. In: Advances in SAR Oceanography from Envisat and ERS Missions, Proceedings of SEASAR, Frascati, Italy, January 23–26, 2006.

17. 17. Bentz, C.M.; Lorenzzetti, J.A.; Kampel, M.; Politano, A.T.; Genovez, P.; Lucca, E.V.D. The contribution of ASTER, CBERS, R99/SIPAM and OrbiSAR-1 data to improve the oceanic monitoring. In: XII Simpósio Brasileiro de Sensoriamento Remoto, Proceedings of XIISBSR, Florianópolis, 2007 [CDROM], Sao Jose dos Campos, Sao Paulo, Brazil, 2007; pp. 3755–3762.

18. 18. Mueller, J.L.; Fargion, G.S.; McClain, C.R. Inherent optical properties - instruments, characterizations, field measurements and data analysis protocols. Ocean optics protocols for satellite ocean color sensor validation, revision 4;. NASA Goddard Space Flight Center: Greenbelt, MA, 2002; NASA/TM-2003-211621/Rev4-Vol IV, p. 76

19. 19. Desa, E.; Suresh, T.; Prabhu Matondkar, S. G.; Desa, E. Sea truth validation of SeaWiFS ocean colour sensor in the coastal waters of the Eastern Arabian Sea. Curr. Sci. 2001, 80, 854-860.

20. 20. Kiefer, D.A. Fluorescence properties of natural phytoplankton populations. Mar. Biol. 1973, 22, 263–269.

21. 21. Loftus, M.E.; Seliger, H.H. Some limitations of the in vivo fluorescence technique. Chesapeake Sci. 1975, 16, 79–92.

22. 22. Dandonneau, Y.; Neveux, J. Diel variations of in vivo fluorescence in the eastern equatorial Pacific: an unvarying pattern. Deep-Sea Res. II 1997, 44, 1869–1880.

23. 23. Hoge, F.E.; Swift, R.N. Absolute tracer dye concentration using airborne laser induced water Raman backscatter. Appl. Opt. 1981, 20, 1191–1202.

24. 24. Reuter, R.; Diebel, D.; Hengstermann, T. Oceanographic laser remote sensing: measurements of hydrographic fronts in the German Bight and in the northern Adriatic sea. Int. J. Remote Sens. 1993, 14, 823–848.

25. 25. Barbini, R.; Colao, F.; Fantoni, R.; Palucci, A.; Ribezzo, S. Differential lidar fluorosensor system used for phytoplankton bloom and seawater quality monitoring in Antarctica. Int. J. Remote Sens. 2001, 22, 369–384.

26. 26. Barbini, R.; Colao, F.; Fantoni, R.; Fiorani, L. Palucci, A. Lidar fluorosensor calibration of the SeaWiFS chlorophyll algorithm in the Ross Sea. Int. J. Remote Sens. 2003, 24, 3205–3218.

Chapter 9

TIME DELAY FOR AERIAL AMMONIA CONCENTRATION MEASUREMENTS IN LIVE-STOCK BUILDINGS

Hans Benny Rom * and Guo-Qiang Zhang

Department of Biosystems Engineering, University of Aarhus, Faculty of Agricultural Sciences

ABSTRACT

Correct measurements of ammonia concentration in air still present considerable challenges. The high water solubility and polarity can cause it to adsorb on surfaces in the entire sampling system, including sampling lines, filters, valves, pumps and instruments, causing substantial measuring errors and time delays. To estimate time delay characteristics of a Photo Acoustic Multi Gas Monitor 1312 and a Multi Point Sampler continuous measurement of aerial ammonia concentrations at different levels was performed. In order to obtain reproducible data, a wind tunnel was used to generate selected concentrations inside and a background concentration representing the air inlet of the tunnel. Four different concentration levels (0.8 ppm, 6.2 ppm, 9.7 ppm and 13.7 ppm) were used in the experiments, with an additional outdoor concentration level as background. The results indicated a substantial time delay when switching between the measuring positions with high and low concentration and *vice versa*. These properties may course serious errors for estimation of e.g. gas emissions whenever more than one measuring channel is applied. To reduce the measurement errors, some suggestions regarding design of the measurement setup and measuring strategies were presented

INTRODUCTION

Correct measurement of ammonia concentrations in air still present considerable challenges, due to the properties of ammonia and state of the art of available technologies. The high water solubility and polarity can cause it to adsorb on surfaces throughout the entire sampling system, including sampling lines, filters, valves, pumps and instruments, with subsequent serious consequences such as time delays, measurement and calculation errors, *etc*. Several measurement systems and principles are described in the literature for detecting aerial ammonia concentrations in the agricultural systems [1]. The instruments have different detection principles and the procedures for doing reliable measurements differ a lot. According to Phillips *et al.* [2] the techniques fall into three categories: detection tubes with a chemical gas specific absorption granulate for instantaneous measurements, accumulative ones that rely on capturing ammonia over time and finally continuous ones equipped with rapidly responding sensors to follow changes in concentration. Detection tubes are relative cheap in purchase but may not be recommendable for measuring concentrations below 2.5 ppm NH3 [2]. The capturing detection techniques such as flux denuders, passive flux samplers and adsorption bottles are primarily used for detection of very low concentrations such as NH3 dry deposition or ambient air concentration levels. The capturing detectors are relatively cheap to purchase, but have high labour costs. Continuous methods like electrochemical cells, chemiluminescence (NOx-analyser), fluorescence, Photo Acoustic Spectroscopy (PAS) and long path optical methods are used for detection of concentration variations such as those found in the measurement of air quality inside livestock buildings and gas emissions from the buildings. The purchase price is high and running costs differ a lot due to different calibration intervals and general maintenance needs. Ni and Heber [1] and Phillips *et al.* [3] have presented details about the agricultural applications and properties of the various systems. During the last decade an increasing number of papers on NH3 emissions have conducted their investigations using PAS instruments such as the Multi-Gas-Monitors and Multipoint Samplers from (Innova) LumaSense Technologies A/S (Denmark). Most of the articles pay little or no attention to data quality parameters such as calibration or quality assessment of the measuring setups used [1,4].

Experiences from use of PAS instruments for measurements of ammonia concentration in air during 15 years show that the instruments need time to indicate correct levels when the ammonia concentration varies from one position to another [5,6]. A number of studies have already focused on ammonia adsorption on various types of tubing materials used in pollutant stream conveyance [7-9]. However, knowledge concerning optimal measurement procedures for a multi-positional data sampling application is missing. Therefore, this paper will focus on the numbers of measurement events or sampling duration and the corresponding accuracy for each position with different ammonia concentration levels to improve the measurement procedures.

MATERIALS AND METHODS

Photo Acoustic Spectroscopy Detection Method

A number of optical filters for specific detection of a large number of different gases can be applied to the Multi-Gas-Monitor [10]. Figure 1 illustrates the measurement principle, which begins with an air sample being drawn into the measurement chamber and the chamber being then sealed by valves. Radiation from an IR-source passes through a chopper and an optical filter into the chamber. The radiation energy is absorbed by the specific gas proportionally to the concentration and subsequently converted to heat, which generates pressure waves. Microphones detect these pressure waves and the signal is post processed and the gas concentration can be estimated. [10]

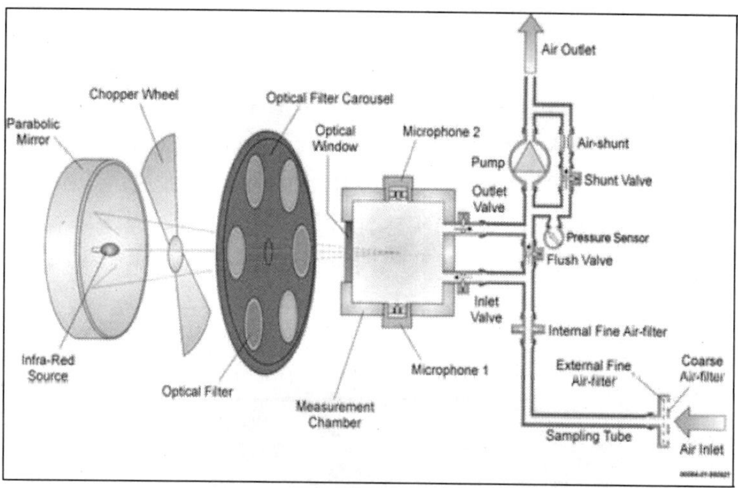

Figure 1. Photo Acoustic Spectroscopy detection principle [10]..

EXPERIMENTAL SETUP

A 1312 Multi-Gas-Monitor 1312 and a 1303 [11] or 1309 [12] Multi Point Sampler were used in the studies. The technical data of the optical filters used for gas detection are listed in Table 1.

Table 1. Filter setup in the Multi Gas Monitor 1312.

Gas	Filter	SIT (s)	Centre wavelength (μm)	Detection threshold (mg/m^3)	Calibration range (mg/m^3)
Carbon Dioxide (CO_2)	UA0982	5	14.1	1.5	1.5–800 1.5–3,500
Dinitrogen oxide (N_2O)	UA0985	5	4.5	0.03	0.03–5.45 0.03–50.0
Ammonia (NH_3)	UA0936	60	9.8	0.2	0.2–74.4
Freon ($C_2H_2F_4$)	UA0972	5	8.8	0.05	0.05–51.6
Methane (CH_4)	UA0969	5	8.0	0.4	0.4–45.1 0.4–203.0

Osada *et al.* concluded that the Multi Gas Monitor overestimates the CO_2, N_2O and CH_4 levels when measurements were carried out in the lower end of the calibration range [13]. In order to improve the accuracy for monitoring both high or low concentrations the optical CO_2, N_2O and CH4 filters were calibrated in two different levels and the calibration data were loaded in two different banks in the monitor. As this was not the case for NH3 and Freon measurements, the double calibration was only done for the CO2, N2O and CH4 filters. The Monitor was configured with a chamber flush time of 60 s and Sample Integration Time (SIT) of 5 s for all gases, except ammonia which was set at 60 s. Nevertheless certified calibration is not at guarantee for correct measurements, because the whole measurement chain such as sampling tubes, dust filters transportation length and instrument properties regarding individual gases have an influence on the reliability

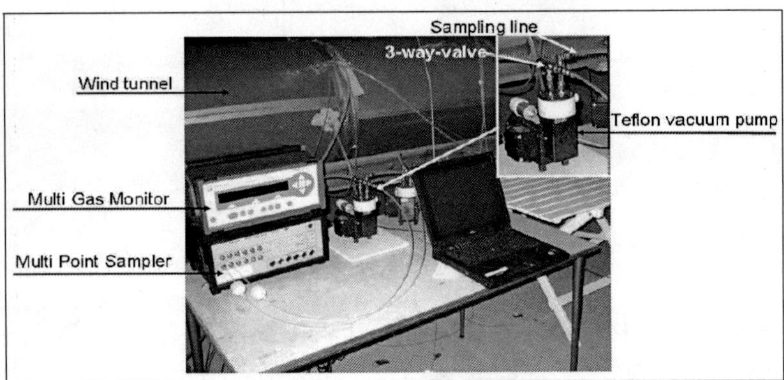

Figure 2. Instrumentation setup.

Figure 2 shows the experimental setup. In order to simulate data collection in practice and to obtain reproducible data a wind tunnel using atmospheric air mixed with clean ammonia was used. The concentration inside the wind tunnel simulates the high concentration position and the outdoor position was used as in the case for data collection in practice. The concentration in the wind tunnel was kept constant during each session of the study. Four sessions were done with ammonia conce ntration levels of 0.8 ppm, 6.2 ppm, 9.7 ppm and 13.7 ppm. The background level was 0.2–0.3 ppm. The Multi

Point Sampler was configured to repeat the measuring event 15 times in the low concentration position before switching to 0.8, 6.2, 9.7 or 13.7 ppm respectively. The cycles of high and low concentration measurements were repeated 14–18 times for concentration ranges up to 9.7 ppm and 4–5 times for 14 ppm. NH3 is often measured simultaneously with CO2, N2O and CH4, which is the reason for including them in this study. Experience shows that among these gases NH3 is the only one displaying a substantial time delay, which is the reason for studying this specific gas. The concentrations of 0.8 ppm and 6.2 ppm in the tunnel were validated by means of three parallel sets of absorption bottles with a strong acid solution. The absorbent was a 0.002 M phosphoric acid (H3PO4) solution and the air flow through the bottles was 2 L/min.

AIR CONVEYANCE FROM THE TUNNEL TO THE INSTRUMENT

A tetraflouroethylene-hexaflouropropylene (FEP) tube of Ø 3/4.8 mm and Ø 6/8 mm (inside/outside diameter) was used for air sampling due to its very low adsorption of ammonia. In order to minimise the time delay caused by the length of the sampling line a type ASF 7015 TF Teflon vacuum pump with bypass (capacity 6–8 L/min) was used to continuously draw the sampling air close to the Multi Point-Sampler-inlet. A three-way connection was applied between each pump and the Multipoint Sampler to ensure the excessive sampled air could exit and thus avoid over-loading the internal pump in the Gas Monitor.

RESULTS AND DISCUSSION

Results

All data were obtained for a minimum of two hours after changing the gas concentration level in the tunnel, allowing the gas concentration in the tunnel to stabilize. The overall results are presented in Figure 3. The data shows the transient approach or time delay when switching between high and low concentration levels and *vice versa*.

According to the analysis of the adsorption in the bottles the average concentration was 0.8 ppm while the 1312 reading was 0.78 ppm after 37.5 min (15 repetitions) in the same position [Figure 3(a)]. According to the 1312 measurements the background was 0.25 ppm after 37.5 min. Due to the very low concentration difference between tunnel and background the calculated error was rather small.

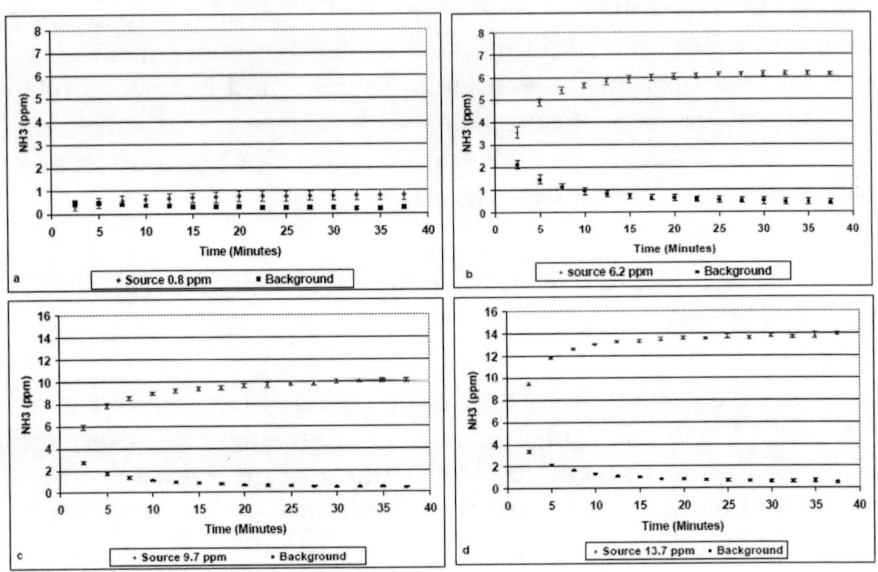

Figure 3. Time delay when switching between high and low concentration positions **(a)** = 0.8 ppm; **(b)** = 6.2 ppm; **(c)** = 9.7 ppm and **(d)** = 13.7 ppm.

By increasing the gas concentration of the air in the tunnel to 6.2 ppm the error increased substantially. The average detection concentration for the 1312 was 6.2 ppm after 37.5 min. showing a substantial error during the first 12.5 to 25 min. [Figure 3(b)]. The similar results were found for the concentrations of 10 ppm [Figure 3(c)] and 14 ppm [Figure 3(d)].

Discussion

The results present substantial error due to rapid change between high and low concentration levels. In order to explain the differences between the measured concentrations and the real concentrations a correction factor was derived. The correction factor to be applied when changing concentration from low to high level was derived according to the following equation:

$$C_f = 1 + \frac{V_t - V_m}{V_t}$$

where C_f = Correction factor; V_t = Target concentration; V_m = Measured concentration.

Figure 4 shows the correction factors derived in this study to be used in order to generate a reliable dataset when changing low to high concentrations. In general measuring periods of 12.5 to 25 minutes are needed in order to obtain reliable data. For example, when switching between 0.3 to 6.2 ppm the correction factors increased from 1.02 to 1.14 when the measuring period was reduced from 25 to 12.5 minutes.

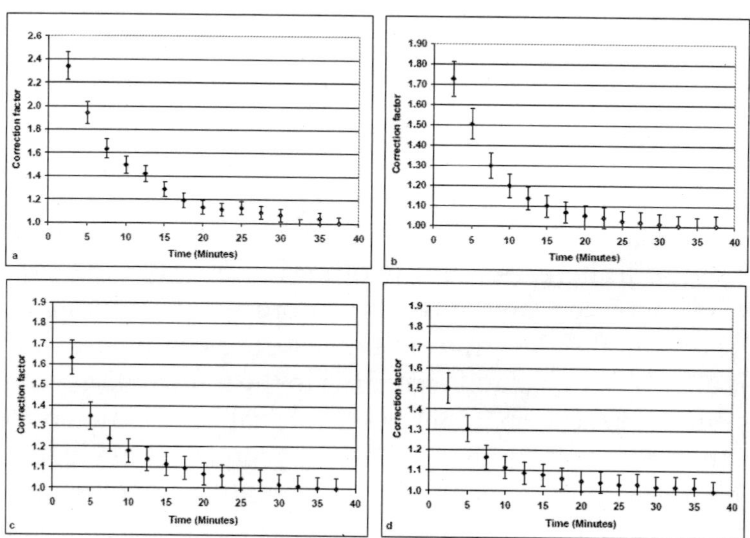

Figure 4. Correction factors with respect to time period in each location **(a)** = 0.8 ppm; **(b)** = 6.2 ppm; **(c)** = 9.7 ppm and **(d)** = 13.7 ppm

There are a number of reasons that could cause the delay. First Ni and Heber [1] and Phillips *et al.* [3] have mentioned that loss of ammonia by adsorption on the inside surfaces of the gas sampling lines and vacuum pumps and by adsorption in condensed water is possible, but these risks were minimized by use of FEP-Teflon and by preventing water condensation by trace heating. Shah *et al.* [8] have studied a number of tube materials. In that study it was demonstrated that only 0.3% of a total flow of 10 ppm NH3 was adsorbed into the FEP-surface. Secondly, if the same tubes were used for sequential sampling of low and high ammonia concentrations, desorption of ammonia from the tubing may result in overestimation of concentration in the low concentration position, at least during the transient period. The opposite is likely to be true when sampling from the high concentration position. In this study separate tubes and vacuum pumps were used for both sampling positions except for the inside surfaces and valves in the Multipoint Sampler and Multiage Monitor and the tube connecting the two units. The total tube length between the Multipoint Sampler and the Multiage Monitor was approximately 0.3 m. In the setup of sampling procedures for the 1312 Multi Gas Monitor it is possible to adjust the flush time for the measuring chamber and for the internal tubes respectively. In this study the flush time was 60 s and 3 s. for the chamber and the tubes. In order to minimize the deviation of the measurements the Sample Integration Time (SIT) can be adjusted from 5 to 60 s for each individual gas. In this case the SIT for the ammonia filter was set to 60 s due to the high water solubility and polarity. The disadvantage of using a longer SIT is that the time required to complete each measurement will increase. On the other hand, shortening the SIT results in an increased deviation, but more measurements can be done. In addition, the design of the measuring chamber could also have some effect on the delay. According to the information provided by the manufacturer the chamber is cylindrical, so it may have some locations that are not properly flushed, consequently causing a memory effect from one event to the next, although this does not seem to be the main reason behind the observed lag, due to the fact that it is mainly NH3 which has the time delay problem and other instruments seem to have almost similar responses to NH3 such as the NOx analyser with an upstream thermal ammonia converter that has a time delay of up to 4 minutes for the 5.4 to 1.1 ppm range

[3]. Unpublished studies carried out in the Climate Laboratory at Institute of Biosystems Engineering, Aarhus University (Denmark), indicated that a NOx-analyser with an internal NH3-NOx converter had a time delay in the same range as for the Multigas Monitor.

The results indicate that the chosen setup required approximately 37-38 minutes (15 repeating measuring events) in each position, depending of the number of gases to be measured and the setup of the monitor. The time delay is mainly due to adherence effects on the surfaces in the instruments and in the tube between the instruments. It is assumed that the main reason is the surfaces and materials inside the instruments, which are used for sequential sampling of low and high ammonia concentrations. To reduce this effect it could be desirable to improve the instrument setup with an option for selection of longer flush periods for the internal tubes, similar to the functions for Chamber flush and SIT.

Measuring ammonia concentration in air seems to be a question of compromises. When measuring only NH3 at various concentration levels, a NOx-analyser with an upstream NH3-NOx-converter could be recommended, but if the aim is to estimate more gases simultaneously then a Multigas Monitor could be an option taking the mentioned disadvantages into account. If the aim is to estimate an average daily level, then passive adsorption methods should be seriously considered.

When doing measurements of ammonia concentrations with substantial variation between various measuring positions, e.g., indoor and outdoor concentrations, it is recommended that a strategy be designed in order to obtain a reliable dataset, or at least develop data for estimation of reliable correction factors taking the consequences of the time delay into account. Big variations of ammonia concentration levels between channels may require longer measuring period in each measurement position and consequently the number of reliable dataset for each position will be reduced. Rapid and big variations in the concentrations may require shorter measuring period in each position, thus increasing dependence on correction factors, which will increase the number of data sets and on the other hand increase the uncertainties for each individual data set.

CONCLUSIONS

In the validation of measurement procedures for aerial ammonia levels using an Innova Multi-Gas- Monitor 1312 equipped with an Innova Multipoint Sampler 1303, the following conclusions were drawn:

1) Substantial time delays occur, when the measuring strategy is designed to switch between high and low concentration positions.

2) It is recommended to plan the measuring strategy to allow a minimum of 12.5 to 25 minutes on each position. Even in this case a correction factor may be needed in order to obtain reliable results for further calculations.

3) It is always recommendable to validate the instrument properties with respect to sensibility to variations in concentration levels before starting any data collection.

REFERENCES AND NOTES

1. Ni, J.Q; Heber, A.J. Sampling and Measurement of Ammonia at Animal Facilities. *Adv. Agr.* **2008**, *98*, 203-257.
2. Phillips, V.R.; Lee, D.S.; Scholtens, R.; Garland, J.A.; neath, R.W. A Review of Methods for measuring Emission Rates of Ammonia from Livestock Buildings and Slurry or Manure Stores, Part 2: monitoring Flux Rates, Concentrations and Airflow Rates, *J. Agric. Eng. Res.* **2001**, *78*, 1-14.
3. Phillips, V.R.; Holden, M.R.; Sneath, R.W.; Short, J.L.; White, R.P.; Hartung, J.; Seedorf, J.; Schroeder, M.; Linkert, K.H.; Pedersen, S.; Takai, H.; Johnsen, J.O.; Groot Koerkamp, P.W.G.; Uenk, G.H.; Scholtens, R.; Metz, J.H.M.; Wathes, C.M. The Development of Robust Methods for Measuring Concentrations and Emission. Rates of Gaseous and Particulate Air Pollutants in Livestock Buildings. *J. Agric. Eng. Res.* **1998**, *70*, 11-24.
4. Ngwabie, N.M.; Jeppsson, K.H.; Nimmermark, S.; Swensson, C.; Gustafsson, G. Multi-Location Measurements of Greenhouse Gases and Emission Rates of Methane and Ammonia from a Naturally-Ventilated Barn for Dairy Cows. *J. Biosyst. Eng.* **2009**, *103*, 68-77.
5. Rom, H.B. Evaluation of Photoacoustic Ammonia Detection in livestock Buildings. In *Proceedings of XII CIGR World Congress and AgEng'94 Conference on Agricultural Engineering.* Milan, Italy, 1994; pp. 238-239.
6. Rom, H.B. Ammonia Emission from Pig Confinement Buildings. System Analysis and Measuring Methods. Ph.D. Thesis. The Royal Agricultural and Veterinary University, Copenhagen, Denmark, 1995; p. 144.

7. Mukhtar, S.; Rose, A.; Capareda, S.; Boriack, C.; Lacey, R.; Shaw, B.; Parnell Jr. C. Assessment of Ammonia Adsorption onto Teflon and LDPE Tubing used in Pollutant Stream Conveyance. *Agri. Eng. Int.: CIGR J. Sci. Res. Develop.* **2003**, *5*, 03-012.

8. Shah, S.B.; Grabow, G.L.; Westerman, P.W. Ammonia Adsorption in five Types of Flexible Tubing Materials. *ASABE App. Eng. Agr.* **2006**, *22*, 919-923.

9. Rose, A.J.; Mukhtar, S.; Lacey, R.E.; Shaw, B.W.; Parnell, C.B. Assessment of Ammonia Adsorption to Polymer Tubings. *ASAE Annual Meeting July* **2003**, p 034106.

10. Luma Sense. Available online: www.lumasense.dk/INNOVA-1412.gas_monitoring4.0.html/ (accessed on 3 April 2010)

11. Luma Sense. Available online: www.lumasense.dk/INNOVA-1303.gas_monitoring6.0.html/ (accessed on 3 April 2010).

12. Luma Sense. Available online: www.lumasense.dk/INNOVA-1309.gas_monitoring5.0.html/ (accessed on 6 April 2010)

13. Osada, T.; Rom, H.B.; Dahl, P. Continuous Measurement of Nitrous Oxide and Methane Emission in Pig Units by Infrared Photoacoustic Detection. *Trans. ASAE* **1998**, *41*, 1109-1114.

Chapter 10

REMOTE WATER TEMPERATURE MEASUREMENTS BASED ON BRILLOUIN SCATTERING WITH A FREQUENCY DOUBLED PULSED YB:DOPED FIBER AMPLIFIER

Kai Schorstein, Alexandru Popescu, Marco Göbel and Thomas Walther *

TU Darmstadt, Institute for Applied Physics, Schlossgartenstr. 7, D-64289 Darmstadt, Germany;

ABSTRACT

Temperature profiles of the ocean are of interest for weather forecasts, climate studies and oceanography in general. Currently, mostly *in situ* techniques such as fixed buoys or bathythermographs deliver oceanic temperature profiles. A LIDAR method based on Brillouin scattering is an attractive alternative for remote sensing of such water temperature profiles. It makes it possible to deliver cost-effective on-line data covering an extended region of the ocean.

The temperature measurement is based on spontaneous Brillouin scattering in water. In this contribution, we present the first water temperature measurements using a Yb:doped pulsed fiber amplifier. The fiber amplifier is a custom designed device which can be operated in a vibrational environment while emitting narrow bandwidth laser pulses. The device shows promising performance and demonstrates the feasibility of this approach. Furthermore, the current status of the receiver is briefly discussed; it is based on an excited state Faraday anomalous dispersion optical filter.

INTRODUCTION

The knowledge of temperature profiles of the upper-ocean mixed layer is highly relevant to oceanography. Furthermore, this data can provide valuable input for climate studies and weather forecasts. Currently, the measurement of temperature profiles is based on *in-situ* techniques such as fixed buoys or bathythermographs, which are deployed from aircrafts or ships. However, these techniques cannot deliver cost-effective on-line data from an extended region of the ocean. The exploitation of spontaneous Brillouin scattering for measuring sound speed and temperature in water was proposed by Hirschberg *et al.* [1] and later demonstrated by Guagliardo *et al.* using pulsed lasers [2] and Hirschberg *et al.* using cw-lasers [3]. In combination with remote sensing lidar techniques, a powerful instrument can be obtained that provides a much more efficient measurement technique compared to the established *in situ* techniques. In general, the lidar working principle is straight forward and is illustrated in Figure 1. A laser pulse is sent into the water. Brillouin scattering occurs in the water and the Brillouin backscattered light is analyzed by a detector unit. Depth resolution is achieved by analyzing the light for the return time. The temperature information is extracted from the spectral content of the signal. Thus, by careful analysis of the backscattered light for time-of-flight and frequency, the temperature profile in the ocean can be measured.

Figure 1: Vision of the Brillouin lidar mounted under a helicopter. The green ray depicts the laser pulse being sent into the ocean. The red and blue rays represent the two Brillouin shifted parts of the backscattered light whose frequencies are slightly red- and blue shifted with respect to the laser frequency.

Specifically, light propagating in water is scattered off moving density fluctuations; due to the Doppler effect the spectrum of the backscattered light shows components shifted to the red and blue of the original light frequency. Thus, the method is sensitive to the local velocity of sound. The shifted frequency components are referred to as Brillouin lines. Frequency shifts for 532 nm-light for typical water temperatures and salinities found in the oceans are in the order of 7-8 GHz [4]. As the temperature dependence of the shift is known, the temperature information can be extracted from the data. The relation between the Brillouin shift v_B and the velocity of sound v_S for backscattered light is given by [2]

$$\nu_B = \pm 2 \frac{n(S,T,\lambda)\, v_s(S,T)}{\lambda} . \qquad (1)$$

The dependence of the refractive index n(S,T,λ) and of the sound velocity vS(S,T) on the salinity S, the temperature T and the wavelength λ are well known [4]. Relying on historical data for the salinity S and measuring the Brillouin shift vB enables us to deduce the temperature T. It should be noted that there is also a dependence on pressure, which has been omitted for clarity in the above equation, but which could be easily taken into account [4].

Laboratory based measurements by Fry and coworkers have shown the feasibility of this approach [5-7]. Theoretical investigations of the accuracy limitations have been performed [4]. It was found that an uncertainty of 1 MHz in the frequency shift measurement corresponds to an uncertainty in the temperature of 0.06 °C when the salinity is known. Assuming a world-wide constant salinity, averaged over historical data, a temperature accuracy of 0.5 °C is possible with a frequency uncertainty of 4 MHz. The maximum depth is limited by the available laser power, the detector sensitivity, the signalto- noise ratio and the absorption in water. Estimates result in a penetration depth of up to 100 m [8]. The spatial resolution in the water is about a meter for pulse durations of 10 ns. The necessary minimum specifications of the complete sensor system are rather stringent: (1) Since operation from a mobile platform is intended, the entire sensor has to be compact, insensitive to vibrations and exhibit low power consumption. (2) In order to provide 1 meter depth resolution, the laser pulse must have a temporal width less than about 9 ns. (3) In order to resolve the Brillouin shift the laser source must produce relatively high energy, narrow linewidth pulses close to the Fourier transform limit. (4) The laser radiation should be close to the absorption minimum of water, i.e. between 380 and 550 nm [9]. (5) The receiver unit must exhibit a high light gathering power, and be able to resolve the Brillouin-shift.

As proposed in earlier publications [10-15], a light source which is compatible with the above specification is a pulsed fiber amplifier. The fiber amplifier is composed of a master oscillator (seed) which defines spectral and temporal properties of the light. Then in a power amplifier the light is further amplified. Since there is no need for any resonant optics, the system is intrinsically insensitive to vibrations. Different concepts for sensitive detection of the frequency shift are possible. The most straight forward solution is the use of an edge filter based on a Fabry-Perot interferometer [16], which is perfectly suitable for a laboratory. Carefully built, it possesses the required frequency resolution and the position of the Brillouin shifted lines can be accurately measured, but a Fabry-Perot and interferometers in general do not have the large acceptance angle required for practical use in the field and are also very sensitive to vibrations. Better alternatives are heterodyne detection [17] or edge filter techniques based on molecular or atomic absorption lines. The latter are static devices insensitive to vibrations capable of transforming a frequency shift into a change in transmission. Excited State Anomalous Dispersion Optical Filters (ESFADOF) are such static devices fulfilling the above requirements.

All the Brillouin-spectra measurements which are presented in the following chapter are performed with a scanning Fabry-Perot interferometer. An additional chapter is dedicated to the ESFADOF concept which documents the current status of development. The ESFADOF will be integrated into the whole system in the near future replacing the FPI detector.

THE LIGHT SOURCE

Recent progress in fiber laser and amplifier technology has facilitated the extension to other operational wavelengths apart from the telecom wavelength region. High power systems have been realized by the introduction of large mode area fibers [18]. Fiber lasers and amplifiers have already proven their great versatility for delivering cw or pulsed radiation. Pulse lengths from femto- to nanoseconds have been realized [19-21]. In our application we are using a three stage Yb:doped fiber amplifier, which can be operated between 1020 nm-1100 nm when pumped at 976 nm [22]. This range makes it easier to match the laser with a receiver based on a molecular or atomic edge filter. Compared to a Nd:YAG laser, a fiber amplifier has significant advantages with regard to sensitivity to vibration, efficiency, weight, physical dimensions and operating wavelength region. The lower output energy obtained from the fiber amplifier could be balanced by the increased repetition rate of up to 5 kHz.

Using a seeding technique near Fourier transform limited pulses can be generated [23]. These are required to accurately measure the peak positions of the Brillouin lines. Energy scaling in this operating regime is in general limited by nonlinear effects. The high peak intensities inside the fiber core, the rather long interaction lengths and the small bandwidth promote the occurrence of stimulated Brillouin scattering (SBS) inside the fiber [24]. The onset of SBS can be prevented or at least shifted to higher energies by: (1) increasing the diameter of the fiber core thus reducing peak intensity, (2) reducing fiber length, (3) decreasing pulse length and (4) increasing bandwidth. Clearly, (3) and (4) are not options for our application. The fiber amplifier which is presented in this contribution has been optimized by using a highly doped and thus relatively short fiber as the third amplifier stage.

Experimental Setup

The experimental setup is shown in Figure 2 and consists of three main parts: the pulse generation unit, the amplifier itself and a second harmonic generation unit. The pulse generation unit consists of a small bandwidth external cavity diode laser (ECDL) that emits cw radiation at the desired wavelength. The diode is protected against back-reflections from the fiber amplifier by two Faraday isolators (FI). An electro-optic modulator (EOM) chops the cw-beam into laser pulses. At this point, the pulse duration and repetition rate are defined. Pulse durations between 10 ns and 1 μs at repetition rates of up to 5 kHz can be realized with our setup. The fiber amplifier can also be operated continuous wave (cw) depending on the position of the analyzer (A) behind the EOM. For adjustment, characterization and optimization of the receiver cw operation is advantageous.

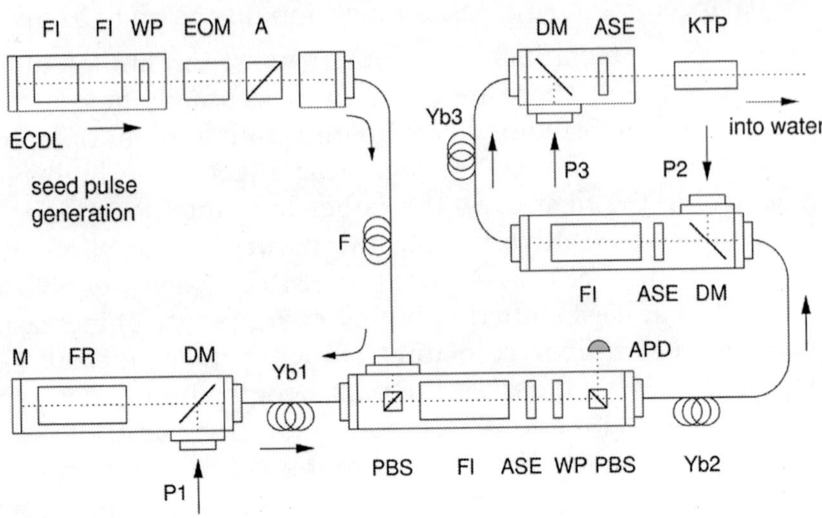

Figure 2 Setup of the fiber amplifier: Yb-doped fibers (Yb1-3), seed laser (ECDL), patch fiber (F), pump lasers (P1-3), electro-optic modulator (EOM), analyzer (A), polarizing beam splitter (PBS), mirror (M), Faraday isolator (FI), Faraday rotator (FR), $\lambda/2$-wave plate (WP), dichroic mirror (DM), interference filters (ASE), avalanche photo diode (APD) and nonlinear KTP crystal (KTP).

Table 1 Fiber properties of the three stages of the fiber amplifier. NA is the numerical aperture.

stage	1	2	3
core diameter	4.4 μm	28 μm	55 μm
inner cladding dia.	-	400 μm	400 μm
NA core	0.15	0.06	0.22
NA inner cladding	-	0.38	0.38
length	0.88 m	12.5 m	0.67 m
Yb concentration	6500 ppm	500 ppm	6500 ppm

The fiber amplifier is an advanced version of the amplifier described in ref. [8, 12, 21] and consists of three consecutive stages. Each stage uses its specific fiber with different geometry, length and dopant concentration. The properties of each fiber are listed in Table 1. Furthermore, the fiber ends are connectorized with high power, air-spaced SMA connectors and polished at an 8° angle to suppress spurious lasing. The first stage consists of a single-mode single-clad fiber while the second and third stage use a double-clad fiber with a multi-mode core.

The amplifier works as follows: A seed pulse arrives at the fiber amplifier and is reflected by the polarizing beam splitter (PBS) into the first stage (Yb1). After a first pass, the amplified pulse is reflected by a Faraday mirror (FR, M) and passes the fiber a second time. The Faraday isolator (FI) protects the first stage from unwanted back-propagating light. The second stage is passed only once. Again, a Faraday isolator (FI) separates the second and the third stage. After a final single-pass amplification in the third stage the high energy laser pulses will be frequency doubled in a KTP crystal (KTP). Interference band pass filters (ASE) are located at certain places to cut out unwanted amplified spontaneous emission.

RESULTS

The pulsed fiber amplifier was operated at 1064 nm with a pulse length of 10 ns. Different parameters affect the overall performance of the fiber amplifier: (1) The polarization of each stage has to be carefully controlled to reduce the loss at the Faraday isolators between the stages and to enhance the efficiency of the second harmonic generation after the third stage. (2) The beam profile has to be carefully adjusted since it is crucial for second harmonic generation efficiency. Both parameters can be controlled by carefully adjusting the coupling into the fiber. In general, it is favorably to excite only a few modes of the fiber. Each mode carries its own polarization and may contribute to a disturbed beam profile with inhomogeneous polarization.

The following characterization was done at a repetition rate of 1 kHz. The seed pulse energy which is launched into the first stage is estimated from the cw peak power and the temporal pulse shape to be 264 pJ. After the first stage a pulse energy of 0.86 μJ is measured, corresponding to a gain of 35.1 dB. The cw pump power of the first stage was approximately 300 mW. No signs of stimulated Brillouin scattering (SBS) or amplified spontaneous emission (ASE) are detectable. After the second stage up to 35.2 μJ are obtained, corresponding to a gain of 16.1 dB with a pump power of 6.4 W. This energy marks the SBS threshold. For stability reasons the second stage was operated below that threshold. Pulse energies of 34.6 μJ at 532 nm are obtained for effective pump powers of 5.7 W and 21.6 W at the second and third stage respectively.

For the actual measurements of the Brillouin spectra which are presented in the following chapter the fiber amplifier was operated well below the SBS threshold in order to improve stability and to avoid the risk of damaging the fibers. The Brillouin-spectrum measurements were performed at a repetition rate of 4.5 kHz with 17.7 μJ at 532 nm, which could be sustained for over 9 hours. The bandwidths of the laser pulses were measured with a scanning Fabry-Perot to be 79.9 MHz; this is within a factor of two of the Fourier transform limit and is similar to prior experiments [23].

BRILLOUIN SCATTERING TEST FACILITY

A water filled tube was used to evaluate the sensor system and to perform water temperature measurements. It consists of two segments divided by glass panes. These are inclined at a small angle in order to suppress back-reflections that disturb the measurements. Each segment can be stabilized to a different water temperature. The water tank is operated with double distilled water for which the salinity is virtually zero. Measurements with increased salinity are straightforward, since all mechanical parts are salt water resistant. The setup consists of the water tube as well as the transmission and receiving optics (cf. Figure 3). The light produced by the fiber amplifier is injected into the water by the mirror M1. The backscattered light is collected at a small angle to the backscattering direction (180°) by the mirror M2 and is directed through the receiving telescope. A pinhole (P) which is placed in the back focal plane of the first lens selects the effective collection volume inside the water tube. Then, the light is directed to the scanning Fabry-Perot interferometer (FPI). Finally, a photomultiplier (PMT) detects the light emerging from the FPI. Clearly, the current configuration is not able to measure range resolved Brillouin spectra, but spectra for a fixed distance can be obtained. The raw spectra taken from single FPI scans can be processed in order to extract the frequency shift information.

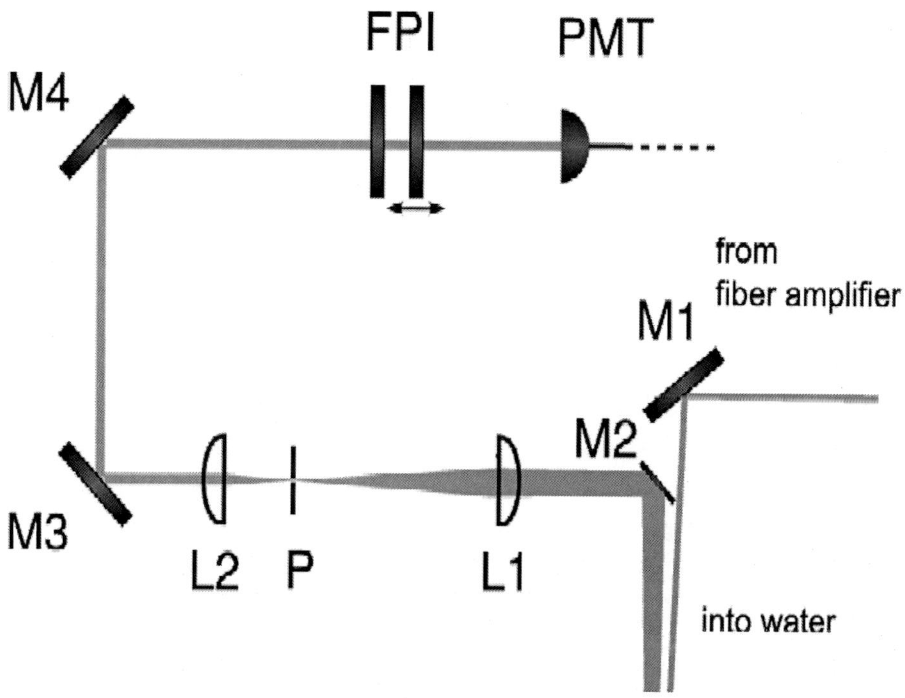

Figure 3 The transmission and receiving optics are composed of mirrors (M), lenses (L), pinhole (P), scanning Fabry-Perot interferometer (FPI) and a photomultiplier tube (PMT).

Since the free spectral range (FSR) of the FPI must be known in order to accurately determine the Brillouin shift, the following calibration procedure has been implemented.

After all measurements have been performed with the pulsed fiber amplifier, the fiber amplifier is switched to the cw operation mode by rotating the analyzer behind the EOM. By rotating the mirror M2 at the receiving optics the cw light is directly sent through the telescope into the FPI. Then, the external cavity diode laser (ECDL) is scanned and resonance peaks can be observed with a separation of the FSR. Simultaneously resonance peaks are recorded from another Fabry-Perot interferometer with a known FSR of 1 GHz. By comparison of these two sets of frequency markers the FSR can be calibrated to yield 20.622(1) GHz.

This is done by exploiting the symmetry of the Brillouin shift with respect to the Rayleigh peak; a quadratic function is found that can be used to correct the nonlinearity of the scan. The measurement of a single spectrum takes 50 seconds. Figure 4 shows a measured and calibrated spectrum recorded for a water temperature of 33.6 ±0.2 °C; it is composed of about 140,000 laser pulses. The Airy peak corresponding to the Rayleigh backscattered component appears twice, once in each order of the Fabry-Perot as it is scanned through two successive orders.

The four smaller peaks originate from the Brillouin backscattered light; they are symmetrically located on either side of their respective Rayleigh peaks. A moving average of 10 was applied to smooth the data. A Levenberg- Marquardt algorithm was implemented to provide a nonlinear least-square fit to these data using an Airy function for the Rayleigh peaks and Lorentz functions for the Brillouin lines. The Brillouin frequency shifts can be obtained from these fits. The fitted function in Figure 4 is indicated by the red curve. The Rayleigh peaks are relatively large, since we have made no effort to remove hydrosols from the distilled water; in fact, we have not changed the water during the past year.

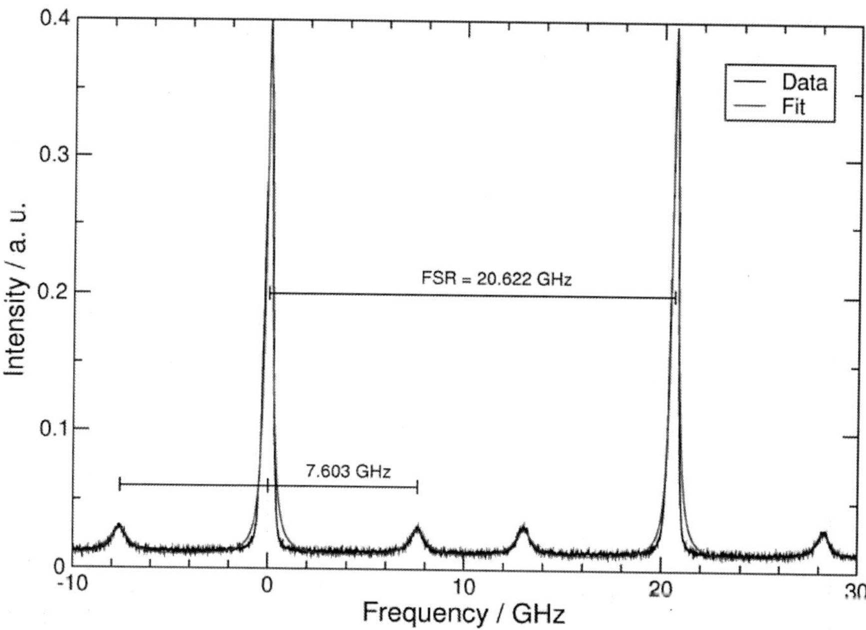

Figure 4 Measured Brillouin spectrum for a water temperature of 33.6 ±0.2 °C. The frequency shift resulting from the fitting procedure is 7.603 ±0.015 GHz.

The results for a series of different water temperatures are compiled in Figure 5. Except for some outliers the agreement between the measured and theoretically calculated [4] frequency shifts is excellent. The outliers are explained by systematic problems with single scans, which could be due to energy fluctuations of the fiber amplifier. The systematic deviation of the data below 15 °C is due to problems achieving a homogeneous temperature distribution in the water tank. Uncertainties in the measurement of the position of the Brillouin peaks directly affect the temperature resolution. When the salinity is zero, a position uncertainty of 1 MHz corresponds to a temperature resolution of 0.04 °C [4].

Converting the mean error of all measurements, 14.4 MHz, into an uncertainty in temperature results in an accuracy of 0.57 °C for the current setup. The following improvements are about to be implemented in the near future in order to increase the temperature accuracy: (1) The energy of each pulse will be monitored and used to correct the data. (2) The stability of the FPI will be improved to increase the frequency resolution. (3) The temperature stabilization of the water tank will be improved to minimize systematics.

Although, our performed measurements are fully equivalent to previous measurements by Fry *et al.* [5, 25, 26], it has to be pointed out that these are the first measurements with the proposed fiber amplifier and constitute an important milestone on the way to a practical implementation of the whole LIDAR system [14]. Despite stability limitations of our FPI, only 17.7 μJ of 532 nm radiation is sufficient to measure the Brillouin shift accurately.

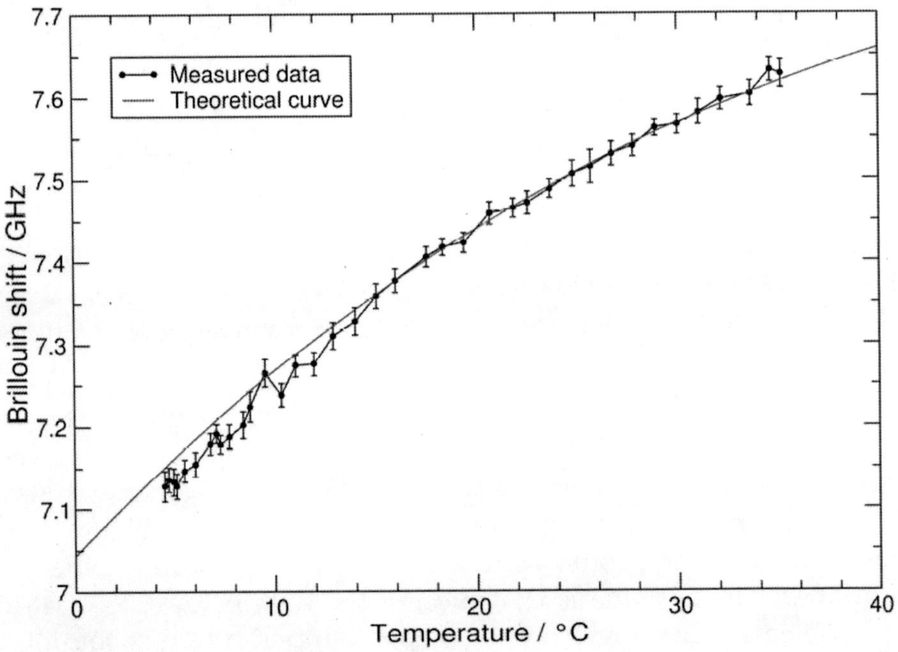

Figure 5 Measurement of the frequency shift versus the water temperature. The red curve indicates the theoretical frequency shift at zero salinity after Fry *et al.* [2].

RECEIVER

The measurement of the Brillouin shifts will be based on a static edge-filter, capable of transforming these small shifts in frequency into large changes of the transmitted intensity. The design criteria of our edge filters can be summarized as follows: (1) Steep transmission edges, (2) symmetrically located in the spectral region of interest, i.e. ±7-8 GHz around the working wavelength, and (3) a maximum overall transmission with a sufficient signal-noise ratio. The symmetry significantly reduces the influence of laser frequency jitter. Several approaches are possible: Fabry- Perot etalons [16], molecular absorption lines [27, 28] or Excited State Faraday Anomalous Dispersion Optical Filter (ESFADOF). Although used in the present study, the Fabry-Perot is simply unsuitable for a practical implementation due to its very small angular acceptance. Molecular absorption lines have limited flexibility, especially when trying to use them as edge filters for both of the Brillouin shifted components simultaneously. The ESFADOF has a large angular acceptance provided by molecular absorption lines but also has the advantage of tunability for the positions of the filter edges.

The ESFADOF is the approach that we are following. For completeness, we will briefly discuss the key ideas, for a more detailed characterization and discussion we refer to our earlier publications [10-12, 14, 15]. Briefly speaking Faraday anomalous dispersion optical filters (FADOFs) and their excited state counterparts ESFADOFs consist of an atomic vapor cell placed in a homogeneous magnetic field between two crossed polarizers. The working principle is as follows (cf. Figure 6): The incident light is polarized by the first polarizer and passes the gas cell. The high anomalous dispersion of the atomic vapor in the vicinity of its absorption lines results in a wavelength dependent rotation of the polarization vector. Finally, the second polarizer, which is rotated by 90o with respect to the first one, converts the rotation of the polarization into a change in the transmitted intensity. As ESFADOFs exploit atomic transitions between two excited states, an external pump source populating the lower ESFADOF state is necessary. Specifically, we investigated the 5P3/2 → 8D5/2 transition in Rubidium as the ESFADOF transition [12, 14]. Our results indicate that an edge filter based on this transition is possible. In order to push the edges to the required separation a magnetic field of BZ=500 mT is necessary. Currently, we are implementing a new cell design which will enable us to apply a field of this magnitude.

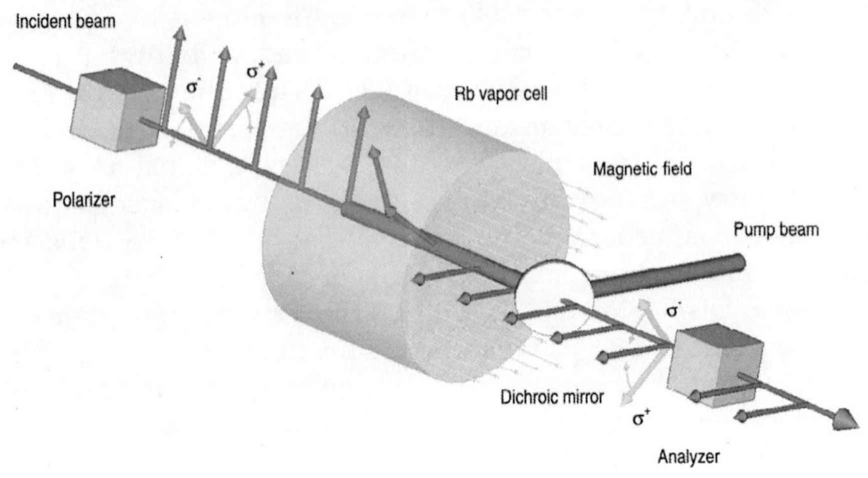

Figure 6 Schematic drawing of an ESFADOF consisting of an atomic vapor cell placed between two crossed polarizers. Due to the presence of the magnetic field, left and right circularly polarized light, σ- and σ+, experience different dispersion in the vicinity of the atomic absorption lines. This results in a significant phase difference between the two polarization components. Therefore, a wavelength dependent rotation of the polarization occurs. The second polarizer projects the rotated polarization onto its basis rotated by 90° with respect to the first one. Therefore an ESFADOF delivers in addition a suppression of daylight.

CONCLUSIONS AND OUTLOOK

We have presented for the first time data proving that accurate measurement of the water temperature by Brillouin scattering employing a frequency doubled fiber amplifier is possible. This is particularly challenging as the need for Fourier transform limited pulses reduces the threshold of stimulated Brillouin scattering in the fibers. Consequently, the pulse energy is limited to much lower values compared to those reported in the literature with much wider bandwidth pulses. Our temperature results agree well with previous measurements. The ESFADOF fulfills the basic requirements as a sensitive detection unit. The next steps will be depth resolved measurements as well as improvements to the ESFADOF, i.e.

pushing the position of the edges to a separation suitable for actual Brillouin measurements.

ACKNOWLEDGMENTS

The authors would like to thank E. S. Fry for valuable discussions, Mathias Sinther for his valuable contributions and Klaus Mörl (IPHT) for providing some of the doped fibers employed in this project. This work is supported in part by the state of Hessen and the DFG within the framework of the Graduiertenkolleg GRK 1114.

REFERENCES

1. Hirschberg, J.; Wouters, A.; Simon, K.; Byrne, J.; Deverdun, C. Laser application to measure vertical sea temperature and turbidity. *NASA Report CR-144865*, Mar. 1976.
2. Guagliardo, J.; Dufilho, H. Range-resolved Brillouin scattering using a pulsed laser. *Rev. Sci. Instr.* 1980, *51*, 79--81.
3. Hirschberg, J.; Byrne, J.; Wouters, A.; Boynton, G. Speed of sound and temperature in the ocean by Brillouin scattering. *Appl. Opt.*, 1984, *23*, 2624.
4. Fry, E.; Emery, Y.; Quan, X.; Katz, J. Accuracy limitations on Brillouin LIDAR measurements of temperature and sound speed in the ocean. *Appl. Opt.* 1997, *36*, 6887-6894.
5. Emery, Y.; Fry, E. Laboratory development of LIDAR for measurement of sound velocity in the ocean using Brillouin scattering. *Proc. SPIE, Ocean Optics XIII* 1996, *2963*, 210-215.
6. Fry, E.; Katz, J.; Nicolaescu, R.; Walther, T. Remote sensing in the ocean: Measurement of sound speed and temperature. *Proc. SPIE, Ocean Optics XIV* 1998.
7. Fry, E.; Katz, J.; Liu, D.; Walther, T. Temperature dependence of the Brillouin linewidth in water. *J. Mod. Opt.* 2002, *49*, 411-418.
8. Hickman, G.; Harding, J.; Carnes, M.; Pressman, A.; Kattawar, G.; Fry, E. Aircraft laser sensing of sound velocity in water: Brillouin scattering. *Rem. Sens. Environ.* 1991, *36*, 165-178.
9. Pope, R.; Fry, E. Absorption spectrum (380-800 nm) of pure water. II. integrating cavity measurements. *Appl. Opt.* 1997, *36*, 8710-8723.
10. Popescu, A.; Schorstein, K.; Walther, T. A novel approach to a Brillouin-LIDAR for remote sensing of the ocean temperature. *Appl. Phys. B* 2004, *79*, 955-961.
11. Popescu, A.; Walther, T. On the potential of Faraday anomalous dispersion optical filter as highresolution edge filters. *Laser Phys.* 2005, *15*, 55-60.

12. Popescu, A.; Walldorf, D.; Schorstein, K.; Walther, T. On an excited state faraday anomalous dispersion optical filter at moderate pump powers for a Brillouin-LIDAR receiver system. *Opt. Comm.* 2006, *264*, 475-481.

13. Schorstein, K.; Popescu, A.; Scheich, G.; Walther, T.; Fry, E. Towards a Brillouin-LIDAR for remote sensing of the temperature profile in the ocean. *OCEANS 2006*, Sept. 2006; pp. 1-6.

14. Schorstein, K.; Scheich, G.; Popescu, A.; Walther, T.; Fry, E.S. A fiber amplifier and an ESFADOF: Developments for a transceiver in a Brillouin LIDAR. *Laser Phys.* 2007, *17*, 975-982.

15. Popescu, A.; Schorstein, K.; Scheich, G.; Massmann, I.; Fietzek, P.; Walther, T.; Fry, E. S. Key developments for a Brillouin-LIDAR for remote sensing of oceanic temperature profiles. *Proceedings of the 3rd EARSeL Workshop on the Remote Sensing of the Coastal Zone*, 2007.

16. Korb, C.; Gentry, B.; Weng, C. Edge technique: theory and application to the LIDAR measurement of atmospheric wind. *Appl. Opt.* 1992, *31*, 4202.

17. Henderson, S. *Laser Remote Sensing*, chapter Wind Lidar. CRC Taylor & Francis, Boca Raton, 2005.

18. Zenteno, L. High-power double-clad fiber lasers. *J. Lightwave Tech.* 1993, *11*, 1435-1446.

19. Limpert, J.; Schreiber, T.; Clausnitzer, T.; Zöllner, K.; Fuchs, H.; Kley, E.; Zellmer, H.; Tünnermann, A. High-power femtosecond Yb-doped fiber amplifier. *Optics Exp.* 2002, *10*, 628- 638.

20. Galvanauskas, A. Mode-scalable fiber-based chirped pulse amplification systems. *IEEE J. Select.Top. QE* 2001, *7*, 504-516.

21. Limpert, J.; Höfer, S.; Liem, A.; Zellmer, H.; Tünnermann, A.; Knoke, S.; Voelckel, H. 100-W average-power, high-energy nanosecond fiber amplifier. *Appl. Phys. B* 2002, *75*, 477-479.

22. Pask, H.; Carman, R.; Hanna, D.; Tropper, A.; Mackechnie, C.; Barber, P.; Dawes, J. Ytterbiumdoped silica fiber lasers: Versatile sources for the 1-1.2 µm region. *IEEE J. Select. Top. QE* 1995, *1*, 2-13.

23. Nicolaescu, R.; Fry, E.; Walther, T. Generation of near-fourier-transform-limited high-energy pulses in a chain of fiber-bulk amplifiers. *Opt. Lett.* 2001, *26*, 13-15.

24. Smith, R. Optical power handling capacity of low loss optical fibers as determined by stimulated Raman and Brillouin scattering. *Appl. Opt.* 1972, *11*, 2489-2494.

25. Fry, E.; Xiao, G.; Katz, J. Measurement of sound speed and temperature in the ocean via remote sensing of Brillouin scattering. *Proc. Ocean Optics XV* 2002.

26. Walther, T.; Katz, J.; Liu, D.; Wan, X.; Fry, E. Temperature dependence of the Brillouin linewidth. Number 1044 in Ocean Optics XIV. ONR, Kona, 1998.

27. Fry, E.S., Brillouin Lidar Receiver Based on Absorption in I2 and Br2, Physics Department Texas A&M University: College Station, TX, 1992.

28. Fry, E.S.; Walther, Th.; Nicolaescu R.; Katz, J. Remote Sensing in the Ocean: measurement of sound speed and temperature. Number 1043 in Ocean Optics XIV. ONR, Kona, 1998.

Citations

CHAPTER 1

Fay, C.; Doherty, A.R.; Beirne, S.; Collins, F.; Foley, C.; Healy, J.; Kiernan, B.M.; Lee, H.; Maher, D.; Orpen, D.; Phelan, T.; Qiu, Z.; Zhang, K.; Gurrin, C.; Corcoran, B.; O'Connor, N.E.; Smeaton, A.F.; Diamond, D. Remote Real-Time Monitoring of Subsurface Landfill Gas Migration. Sensors 2011, 11, 6603-6628. doi:10.3390/s110706603.

CHAPTER 2

Kubota, M.; Hihara, T. Retrieval of Surface Air Specific Humidity Over the Ocean Using AMSR-E Measurements. Sensors 2008, 8, 8016-8026. doi:10.3390/s8128016.

CHAPTER 3

Zhang, C.; Chen, G. A First Comparison of Simultaneous Sea Level Measurements from Envisat, GFO, Jason-1, and TOPEX/Poseidon. Sensors 2006, 6, 235-248.

CHAPTER 4

Koch, M.; Wiese, M. Quality Visualization of Microarray Datasets Using Circos. Microarrays 2012, 1, 84-94. doi:10.3390/microarrays1020084.

CHAPTER 5

Le Borgne, Y.-A.; Raybaud, S.; Bontempi, G. Distributed Principal Component Analysis for Wireless Sensor Networks. Sensors 2008, 8, 4821-4850. doi:10.3390/s8084821.

CHAPTER 6

Jehle, M.; Perler, D.; Small, D.; Schubert, A.; Meier, E. Estimation of Atmospheric Path Delays in TerraSAR-X Data using Models vs. Measurements. Sensors2008, 8, 8479-8491. doi:10.3390/s8128479.

CHAPTER 7

Fanise P, Pardé M, Zribi M, Dechambre M, Caudoux C. Analysis of RFI Identification and Mitigation in CAROLS Radiometer Data Using a Hardware Spectrum Analyser. Sensors. 2011; 11(3):3037-3050.

CHAPTER 8

Kampel, M.; Lorenzzetti, J.A.; Bentz, C.M.; Nunes, R.A.; Paranhos, R.; Rudorff, F.M.; Politano, A.T. Simultaneous Measurements of Chlorophyll Concentration by Lidar, Fluorometry, above-Water Radiometry, and Ocean Color MODIS Images in the Southwestern Atlantic. Sensors 2009, 9, 528-541. doi:10.3390/s90100528.

CHAPTER 9

Hans Benny Rom and Guo-Qiang Zhang, Time Delay for Aerial Ammonia Concentration Measurements in Livestock Buildings, doi:10.3390/s100504634

CHAPTER 10

Kai Schorstein, Alexandru Popescu, Marco Göbel and Thomas Walther, Remote Water Temperature Measurements Based on Brillouin Scattering with a Frequency Doubled Pulsed Yb:doped Fiber Amplifier, DOI: 10.3390/s8095820.

INDEX